# Structure and Approximation in Physical Theories

# Structure and Approximation in Physical Theories

Edited by

**A. Hartkämper**

and

**H.-J. Schmidt**

*University of Osnabrück*
*Osnabrück, FRG*

**PLENUM PRESS • NEW YORK AND LONDON**

Library of Congress Cataloging in Publication Data

Colloquim on Structure and Approximation in Physical Theories (1980 : Osnabrück, Germany)
Structure and approximation in physical theories.

"Proceedings of a Colloquium on Structure and Approximation in Physical Theories, held at Osnabrück, FRG, in June 1980"–T.p. verso.
Bibliography: p.
Includes index.
1. Physics–Methodology–Congresses. 2. Approximation theory–Congresses. I. Hartkämper, A. II. Schmidt, H.-J. (Heinz-Jürgen), 1948- . III. Title.
QC5.56.C66 1980 530'.028 81-15846
AACR2

ISBN-13:978-1-4684-4111-6 e-ISBN-13:978-1-4684-4109-3
DOI: 10.1007/978-1-4684-4109-3

Proceedings of a colloquium on Structure and Approximation in Physical Theories held at Osnabrück, FRG, in June 1980


Softcover reprint of the hardcover 1st edition 1981
A Division of Plenum Publishing Corporation
233 Spring Street, New York, N.Y. 10013

## PREFACE

The present volume contains 14 contributions presented at a colloquium on "Structure and Approximation in Physical Theories" held at Osnabrück in June 1980. The articles are presented in the revised form written after the colloquium and hence also take account of the results of the discussion at the colloquium.

It is a striking feature that the problem of approximation in physical theories has only recently found some attention in the philosophy of science, although the working physicist is constantly confronted with those questions. No interesting theory of exact science exactly fits its experimental data; almost every relation between different theories is an approximate one. Therefore an adequate reconstruction of physical theories must take into account and conceptualize the moment of approximation. The majority of the articles in this book is centered around this subject.

There are at least two elaborate, 'structuralistic' approaches to the formalization of physical theories in which the aspect of approximation has been incorporated: the approach due to P. Suppes, J. Sneed, W. Stegmüller ("S-approach") and the approach of G. Ludwig and his co-workers ("L-approach"). The articles in this book correspondingly fall into three classes: presentation, elaboration and critique of the L-approach [Hartkämper/Schmidt, Ludwig, Neumann, Werner, Schmidt, Mayr, Kamlah, Majer, Grafe] or of the S-approach [Moulines, Balzer, Cooke], and articles referring to both approaches or concerned with related matters [Scheibe, Pfarr, Castrigiano]. Of course, this is only a rough classification and each article must be appraised in its own right.

It is appropriate to indicate among the topics touched on in this book are the following:

1. Discussion of different kinds of imprecision and its mathematical representation by uniform structures [Ludwig, Moulines, Balzer, Mayr, Schmidt].

2. Physical examples of approximation: Kepler-Newton approximation [Moulines, Mayr], approximate quantum dynamics [Werner] and approximate causality [Neumann].

3. Philosophical reflections on the concept of approximation [Majer, Grafe, Balzer].

4. Comparison of the formal apparatuses of the S- and L-approach [Scheibe].

5. Axiomatizations of special physical theories: Geometry and the theory of grinding plates [Castrigiano], Special relativity [Pfarr], Classical particle mechanics [Cooke].

ACKNOWLEDGEMENT

The editors gratefully acknowledge the sponsorship of the Deutsche Forschungsgemeinschaft concerning the colloquium, the efforts of the authors, the cooperation of Plenum Publishing Corp. in publishing these proceedings and the kind assistance of Dr. R. Cooke and Dipl.-Phys. R. Werner concerning the editing and the final linguistic form of the manuscripts. Last not least we would like to thank Frau P. Ellrich and Frau A. Schmidt for retyping the manuscripts.

A. Hartkämper
H.-J. Schmidt

May 1981
University of Osnabrück

CONTENTS

# A SHORT ACCOUNT OF THE L-PROGRAM

A. Hartkämper and H.-J. Schmidt

Fachbereich Physik der Universität
Postfach 4469/ 4500 Osnabrück
Federal Republic of Germany

## I.

As pointed out in the preface, this volume will be concerned for the greater part with the structure of physical theories and the role of approximation with respect to two programs of theory reconstruction: the "S-program" (P. Suppes, J. Sneed, W. Stegmüller) and the "L-program" (G. Ludwig). The reader who is not acquainted with the S-program can consult some introductory literature [1,2,3] whereas Ludwig's work [4-11] has not yet been published in English. Therefore a short account of the L-program might be useful for the non-German reader to understand those articles which deal with this program. It should be mentioned that some articles in this book (esp. Kamlah, Scheibe,Majer) give a more detailed introduction to certain aspects of the L-program and may round off our account. (Articles will be cited by the author's name in brackets.) The authors have been requested to use the standardized translation of Ludwig's technical terms which will be established in this account. Exceptions to this standard terminology as well as some remarks on the notation used in this volume[4)] will be indicated in the footnotes.

## II.

When discussing the L-program one should bear in mind that Ludwig's approach to the basic structures of a physical theory has not been set up in the context of the contemporary discussion in philosophy of science and methodology. It has rather been created as a framework to cope with physical problems, e.g. the interpretation of quantum mechanics and the relation between macrophysics and microphysics. The contribution of working physicists to philosophy of science has a long tradition: we only mention the work of v. Helmholtz, N. Bohr and Bridgman. It is therefore not surprising, that the L-approach cannot easily be classified within the current "...isms" considered in philosophy of science. It is more appropriate to lay emphasis on the physical motivations when giving an account of this approach.

To the layman, physics appears as a complex human enterprise consisting of different actions and interactions of physicists: constructing and operating apparatuses (preparing and measuring), symbolic manipulations (mathematical deductions, numerical calculations) and communication. According to what physicists themselves say, the reason of this enterprise is to construct and to apply *physical theories* ($PT$'s). Leaving aside the question of the ultimate goal of the whole buisiness, we may ask: what is a $PT$, or, less pretentiously, how does a $PT$ work? According to the L-conception of theories a $PT$ "pictures", or represents, certain elements of reality via corresponding elements in a mathematical theory. In Ludwig's terms: there is a *correspondence* (——) between a *mathematical theory* $MT$ and a *domain of reality* [Wirklichkeitsbereich] $W$ described by the particular $PT$. The correspondence (——) is not an exact mapping <—> but a blurred one, speaking metaphorically. An element of $MT$ corresponds to *different* physical situations, and vice versa. These explanations, abbreviated as $PT = W$ (——) $MT$, can be viewed as a (provisional) definition only if the components $W$, (——), $MT$ of the definiens could be determined independently of $PT$. This entails regarding $MT$ as a formal theory without "built-in" physical interpretation. This is for instance done in the work of N. Bourbaki, on which Ludwig rests his approach. But the objects of $W$ the $PT$ is talking about, e.g. the fields $\vec{E},\vec{B}$ in electrodynamics, are typically defined *within* $PT$. This poses a problem, sometimes called the "problem of theoretical terms", which seems to be a point of bi-furcation, or rather multi-furcation, for the different approaches towards understanding physical theories (cf. Kamlah, Cooke). Ludwig's solution goes as follows: He restricts the domain of reality $W$ to a *basic domain of applications* [Grundbereich] $G$, such that (1) $G$ (——) $MT$ may be defined independently of $PT$ and (2) $W$ may be *reconstructed* in terms of $G$ (——) $MT$. This policy could be described as some kind of "cautious realism": a term of a $PT$ is not supposed to picture an element of $W$ unless it is reconstructed from $G$, of course using the "laws" of $PT$. This implicit appeal of Ludwig to precaution becomes significant if one recalls some premature realistic interpretations of e.g. certain terms in quantum theory like "the collapse of the wave function". This prominent "collapse" disappears if it is traced back to the basic domain $G$ of quantum theory (consisting of macroscopic apparatuses).

At this point, several questions arise.

(1) What distinguishes $G$ from $W$?

(2) How can we exclude that the reconstruction of $W$ suffers

from (a) $PT$ being a 'bad' theory or (b) $PT$ being a 'poor' theory?

The answers given by Ludwig require a more detailed account of the formalism underlying the $G$ (—) $MT$ relation.

The first question is answered by the proposal that $G_i$ of $PT_i$ is partially given by means of the $W_{i_1} \ldots W_{i_n}$, $i_1 \neq i$, of other theories $PT_{i_1}$. They are called pre-theories [Vortheorien] of $PT$. Thus a network of theories is suggested which must not contain cycles. At any case, a reference to "directly given facts" described in the every-day language cannot be avoided. The details of the relation pretheory-theory are not yet worked out.

With respect to question (2) we must refer to the metamathematical concept of a species of structures $\Sigma$ (cf. Scheibe, Schmidt). It is defined in such a way, that the usual mathematical theories as e.g. group theory, differential geometry, functional analysis determine certain species of structures. Roughly speaking, $\Sigma$ contains a number of 'abstract' sets, some structures over these sets and some axioms.

Starting within a appropriate species $\Sigma$, a finite number of sets $Q_\nu$ and relations $R_\mu$ is singled out in $MT_\Sigma$ and brought into correspondence with different sorts of physical objects of $G$ and different physical relations between these objects. The $Q_\nu$ are called interpreted sets [Bildmengen], the $R_\mu$ interpreted relations [Bildrelationen][1)] and the rules regulating this correspondence are called the correspondence rules [Abbildungsprinzipien][2)]. The application of the correspondence rules to a physical situation in $G$, called 'Realtext' (collection of facts), yields a number of sentences in $MT_\Sigma$, the observational report $A$ [Abbildungsaxiome][3)]. The first attempt to define a 'good' theory is to ask whether it is consistent with the axioms of $\Sigma$. For example: do the spectral lines of hydrogen coincide with differences of eigenvalues of the Hamiltonian as predicted by quantum theory? Of course, this coincidence is not expected to be an exact one, thus we should weaken the criterion for 'good' theories to the following: is the blurred observational report $\tilde{A}$ consistent with the axioms of $\Sigma$? (The details of "blurring a proposition" by using imprecision-sets are explained in the articles of Ludwig and Moulines.)

There is another obstacle indicated by the notion of a 'poor' theory in question (2). By this we mean that it could happen that an enormous mathematic apparatus $MT_\Sigma$ has very weak empirical consequences, say for instance, $MT_\Sigma$ implies only that the spectral lines of hydrogen form a discrete subset of $\mathbb{R}$. In such a case one would not assign the status of elements of reality to all elements of $MT_\Sigma$.

But since most cases will not be as obvious as this example, the following question arises: Can we find a method for reconstructing physical theories which guarantes that the elements of $MT_\Sigma$

possess physical meaning?

The solution proposed by Ludwig runs as follows: $PT$ has to be formulated in a physically equivalent form such that the basic sets and the structural terms in $\Sigma$ are identical with the interpreted sets and interpreted relations of $PT$. The axioms of $\Sigma$ then represent the physical laws of $PT$ expressed in statements about $G$ (at least if the axioms quantify only over sets constructed in $MT_\Sigma$). This form of a physical theory is called an axiomatic basis of $PT$. The important feature of an axiomatic basis is that mathematical constructions within $MT_\Sigma$ obtain per se a physical meaning. If for instance a new quantity is defined as a theoretical concept within an axiomatic basis, a class of methods for measuring this quantity may be found via the mathematical construction. This exactly mirrors what the physicist does when constructing a measuring device by using the very theory which he is testing. In this way the operationalism of the thirties is revived and simultaneously superseded by a more refined measurement concept.

The concept of an axiomatic basis is also essential for Ludwig's definition of the relation "$PT_2$ is more comprehensive than $PT_1$". It is just this concept which insures that a correspondence at the theoretical level between $PT_1$ and $PT_2$ induces a correspondence at the level of empirical applications.

This intertheoretical relation is usually termed the reduction relation: $PT_2$ reduces $PT_1$. Ludwig splits up this relation into two steps: restriction plus embedding (cf. Mayr, Werner). He also suggested how to incorporate the moment of approximation into these relations.

Consider for example as $PT_2$ the spherical geometry of the earths surface, take its restriction $PT_2'$ to a bounded area of, say $1km^2$. Then $PT_2'$ can be approximately embedded into Euclidean geometry in 2 dimensions $PT_1$.

It remains a task to reconstruct the network of physical theories in their axiomatic basis and to explicate the various relations of pretheory-theory and reduction relations according to the L-program.

NOTES

1) U. Majer uses the translation "picture relation".

2) E. Scheibe uses the translation "empirical interpretation rules".

3) U. Majer uses the translation "axioms of representation".

4) The authors of this volume use a slightly differing notation. This cannot be completely avoided although the editors have tried to standardize the notations as far as possible.

The following symbols will occur frequently:

| | | |
|---|---|---|
| $\mathcal{P}(X)$, Pot(X) | : | the set of all subsets of X |
| $\Rightarrow$, $\Leftrightarrow$ | : | logical implication, equivalence |
| $\subset$ | : | set-theoretical inclusion |
| $\subsetneq$ | : | strict inclusion |
| $\mathbb{R}_+$ | : | the set of reals > 0 |
| $\mathbb{R}_{0+}$ | : | the set of reals ≥ 0 |
| $\forall$, $\wedge$ | : | for all ..., universal quantifier |
| $\exists$, $\vee$ | : | there exists a ..., existential quantifier |

REFERENCES

[1] J.D. Sneed: The Logical Structure of Mathematical Physics, Reidel, Doordrecht 1971.

[2] W. Balzer, J.D. Sneed: Generalized Net Structures of Empirical Theories, Studia Logica 36 (1977) and 37 (1978).

[3] W. Stegmüller: The Structuralistic View of Theories, Springer, Berlin-Heidelberg-New York 1979. (See also literature quoted there.)

[4] G. Ludwig: Deutung des Begriffs "physikalische Theorie" und axiomatische Grundlegung der Hilbertraumstruktur der Quantenmechanik durch Hauptsätze des Messens, Lecture Notes in Phys. 4, Springer, Berlin-Heidelberg-New York 1970. (See chapter II)

[5] G. Ludwig: Einführung in die Grundlagen der theoretischen Physik (4 volumes), Vieweg, Wiesbaden 1974-1978.

[6] G. Ludwig: Die Grundstrukturen einer physikalischen Theorie, Springer, Berlin-Heidelberg-New York 1978.

[7] G. Ludwig: Axiomatische Basis und physikalische Begriffe, in: W. Balzer and A. Kamlah (eds.), Aspekte der physikalischen Begriffsbildung, Vieweg, Wiesbaden 1979.

[8] G. Ludwig: Wie kann man durch Physik etwas von der Wirklichkeit erkennen? Akad. d. Wiss. u.d. Lit., Mainz; Steiner, Wiesbaden 1979.

[9] G. Ludwig: Die Rolle der Mathematik in einer physikalischen Theorie, in: H. Nelkowski et al. (eds.), Einstein Symposion Berlin, Lecture Notes in Phys. 100, Springer, Berlin-Heidelberg-New York 1979.

[10] G. Ludwig: Is the geometry of the real space a form of pure sensible intuition, or a technological construction, or a structure of reality? In: D. Mayr and G. Süßmann (eds.), Space, Time, and Mechanics; Basic Structures of Physical Theories, Reidel, Doordrecht, to appear 1981.

[11] G. Ludwig: Axiomatische Basis einer physikalischen Theorie und theoretische Begriffe, to appear in Zeitschrift für allgemeine Wissenschaftstheorie.

# IMPRECISION IN PHYSICS

G. Ludwig

Fachbereich Physik der Philipps-Universität
Renthof 7
3550 Marburg
Federal Republic of Germany

## I. INTRODUCTION

One of the most fundamental facts of experimental physics a student has to learn at the beginning of his studies is that no measurement is precise. He learns to estimate the margin of error of the measurements he has carried out. What about these imprecisions and so-called errors of measurements?

A theoretical physicist will realize that there is another field of imprecision. He will realize that every mathematical theory used as a picture of reality cannot be regarded as a precise picture. He will realize that there is no difference in principle between a so-called exact theory and an approximation picture of reality. What is the meaning of approximation and imprecision in this field?

In quantum mechanics the so-called uncertainty relation of Heisenberg is under discussion. What is here uncertain? Or does the uncertainty relation describe an imprecision of measurement?

There is another fundamental fact in quantum mechanics: Not every two observables can be measured "together". We want to use the word "together" instead of "simultaneously". The words "measured simultaneously" can be misunderstood. That two observables cannot be measured together has nothing to do with the question at what time the measuring processes which are carried out at different times cannot be carried out "together". Not only the position at a time $t_1$ and the momentum at the same time $t_1$ cannot be measured together, also the position at a time $t_1$ and the position at another time $t_2$ ($\neq t_1$) cannot be measured "together". But we cannot go into details here (see [3] XIII, § 5). We want to discuss the fact that two observables that cannot be measured together can be measured together with imprecisions. What is the meaning of imprecision here?

The four mentioned examples give situations in which words like imprecision, uncertainty, inaccuracy (or similar words) are used. One suspects that very different concepts are indicated with these words. Therefore an analysis of what is meant by imprecision in physics is essential for a philosphy of science.

## II. IMPRECISION BETWEEN THEORY AND REALITY

We shall see later that the first two examples (imprecision of measurements and the imprecision of the connection between a mathematical theory and the physical reality that will be described by the mathematical picture) have an intrinsic connection. The second two examples are of a very different kind. Therefore we shall begin by finding a mathematical description of imprecision as described in the two first examples.

To find and to explain this mathematical description it is profitable to take a special and most simple example of a physical theory. Then it is not difficult to generalize the results. As an example we take the mathematical picture of three dimensional space. Let X be a set, the elements of which will be used as labels for various spots of a "real space", e.g. spots fixed on objects here in this room. In the mathematical theory a distance function $d : X \times X \to \mathbb{R}_{o+}$ is defined. This distance function will be the mathematical picture of distances, measured between the various sponts fixed in this room. Let (X,d) be a Euclidean space.

As theorists we know it is an idealization to consider X as a continuum. Idealization means that it is not possible experimentally to test the idealized structure, e.g. the continuum structure of X. For two points $x_1, x_2 \in X$ with a distance of $10^{-100}$cm there are no distinct facts at all in real space corresponding to these two points. But since we do not know whether there is a lower bound to distances possessing physical significance (and if there is a bound, how small it is) we supersede our ignorance by an idealization namely by the idealization of a continuum.

In our example the space X is endowed with another idealization, namely with the idealized structure of infinite extension. The infinite extension of X is only a substitute for our ignorance regarding the extent of real space.

The fact that we substitute our ignorance by idealized structures in the mathematical theory shows that we do not regard the mathematical theory as a precise picture of reality. There is some imprecision in the relation between a mathematical theory and the reality we want to describe with the theory.

If there is no mathematical theory which is a precise picture of reality, we have to seek a mathematical structure suitable to describe the imprecision in the relation between the mathematical theory and the reality. But since we do not know the magnitude of imprecision we prefer to formulate the "structure of imprecision" in such a way that the "precision" in the mathematical picture can be improved more and more without any finite limit. This last aim is again an idealization, necessary because of our ignorance concerning the finite limit of precision in the relation between mathematical picture and reality. In this sense we try to introduce an idealized "structure of imprecision" which compensates - or better - partially conpensates the idealizations introduced before. What could be a suitable structure of imprecision in the space X, which partially compensates the idealization e.g. of the continuum structure of X?

It is my opinion that what in mathematics is called a uniformity or uniform structure is suitable to describe the imprecision in the interpretation of a physical theory. (A more general formulation of the following considerations is given in [1]).

To make it plausible that a uniformity is suitable for this purpose we use out example of the three dimensional Euclidean space X.

If we use (X,d) as a description of several spots in a room and their measured distances we see that not every two points $x_1$, $x_2 \in X$ used as pictures of two spots can be distinguished as real spots, for instance if the distance of the two spots is less than $10^{-10}$cm. We will try to describe this relation of indistinguishability by a subset $\mathcal{U}$ of $X \times X$. $\mathcal{U}$ may be such that $(x_1,x_2) \in \mathcal{U}$ means that $x_1$, $x_2$ (as representatives of spots) cannot be distinguished. If we make smaller spots and if we measure the distances more precisely we can take a smaller set $\mathcal{U}$ to describe the imprecision in the relation between X and experiments. But if we make the set $\mathcal{U}$ too small we might get a contradiction between theory and experiment since the real imprecisions of measurements might be bigger than described by the set $\mathcal{U}$. We see that the choice of the imprecision set $\mathcal{U}$ depends on the experiments, i.e. on the characterization of the spots and on the methods of measuring distances.

Since we can use several imprecision sets $\mathcal{U}$ according to the various experimental methods, we shall use a subset $\mathcal{N} \subset \mathcal{P}(X \times X)$, the elements $\mathcal{U}$ (sometimes called entourages) of which can be used as imprecision sets. If $\mathcal{U}_1$ is an imprecision set, which makes it possible to compare theory and experiment without contradictions, then also every set $\mathcal{U}_2 \supset \mathcal{U}_1$ is an inprecision set which makes it possible to compare theory and experiment without contradictions. Therefore we postulate for $\mathcal{N}$:

a) $\mathcal{U}_1 \in \mathcal{N},\ \mathcal{U}_2 \supset \mathcal{U}_1 \Rightarrow \mathcal{U}_2 \in \mathcal{N}$.

Since a point x cannot be distinguished from itself we postulate for $\Delta := \{(x,x) \mid x \in X\}$

b) $\Delta \subset \mathcal{U}$ for all $\mathcal{U} \in \mathcal{N}$.

If $x_1$, $x_2$ cannot be distinguished, then also $x_2$, $x_1$ cannot distinguished. Therefore we postulate:

c) $\mathcal{U} \in \mathcal{N} \Rightarrow \mathcal{U}^{-1} \in \mathcal{N}$

where $\mathcal{U}^{-1} = \{(x_1,x_2) \mid (x_2,x_1) \in \mathcal{U}\}$. c) expresses that the relation "$x_1$ and $x_2$ cannot be distinguished" is a symmetrical relation.

The next two axiomatic relations for $\mathcal{N}$ are idealizations describing the situation that we do not know how small the imprecision between the mathematical picture and reality is, or how far the imprecision of measurements can be reduced. Therefore we wish to formulate this ignorance by the idealization that the precision can be improved step by step without limit. This does not mean that we can actually improve experimental precision indefinitely. It indicates only that a limit to experimental precision is unknown to us. The belief in infinitely high precision however is false.

Our ignorance of the finite limit of imprecision is expressed in the following two relations:

d) $\mathcal{U}_1, \mathcal{U}_2 \in \mathcal{N} \Rightarrow \mathcal{U}_1 \cap \mathcal{U}_2 \in \mathcal{N}$.

e) For every $\mathcal{U} \in \mathcal{N}$ there is a $V \in \mathcal{N}$ with $V^2 \subset \mathcal{U}$.

$V^2$ is defined by $V^2 = \{x_1,x_2) \mid$ there is an $x_3$ with $(x_1,x_3) \in V$ and $(x_3,x_2) \in V\}$. e) expresses that for every imprecision set $\mathcal{U}$ there is an other one V which is only "half as big" as $\mathcal{U}$.

$\mathcal{N}$ is said to be separating if $\bigcap_{\mathcal{U} \in \mathcal{N}} \mathcal{U} = \Delta$. The axiomatic relations a) to e) are not complete for such uniform structures which describe physical imprecisions. We have to add a relation which says something about the fact that the refinement of the imprecision described by e) can only be realized step by step, i.e. at every time only by finite steps.

But before we formulate such a relation we want to illustrate the previous relations a) to e) with an example. We choose as an example the three dimensional space X with a distance function d describing the physical space as outlined above. We shall see that the structure $\mathcal{N}$ cannot be found "a priori" but is determined by the physical methods we employ for distinguishing spots in real space.

In our example of the space X we want to use $\mathcal{N}$ to say that very small and very large values of d have no physical significance. To do this introduce the function

$$\varphi(x_1,x_2) = \arctan\left(\frac{d(x_1,x_2)}{D}\right)$$

where D is an arbitrary constant.

Let $N^{(1)}$ be the uniform structure generated by the sets

$$U^{(1)}_{\varepsilon,y_1,\ldots,y_n} = \{(x_1,x_2)\,|\,|\varphi(x_1,y_i)-\varphi(x_2,y_i)|<\varepsilon, i=1,\ldots,n\},$$

where $\varepsilon$ runs through $\mathbb{R}_+$ and $y_1\ldots,y_n$ run through X.

$N^{(1)}$ describes the possibility of distinguishing spots $x_1,x_2$ by the measurement of the distance $d(x_1,y_i)$ and $d(x_2,y_i)$ from fixed spots $y_1,\ldots,y_n$. But this uniform structure $N^{(1)}$ seems to be too coarse, since it is possible to measure directions by telescopes for instances. Therefore we introduce the angle $\chi(x_1,x_2;y)$ between the two directions from y to $x_1$ resp. to $x_2$. $\chi(x_1,x_2;y)$ can be calculated in the mathematical theory of X from the distance function d.

In addition to the entourages $U^{(1)}_{\varepsilon,y_1,\ldots,y_n}$ we introduce the entourages

$$U^{(2)}_{\delta,y_1,\ldots,y_m} = \{(x_1,x_2)\,|\,|\chi(x_1,x_2;y)|<\delta\}.$$

Let $N^{(2)}$ be the uniform structure generated by the entourages $U^{(1)}_{\varepsilon,y_1,\ldots y_n}$ and $U^{(2)}_{\delta,y_1,\ldots,y_m}$. $N^{(2)}$ is finer than $N^{(1)}$.

$(X,N^{(1)})$ is uniformly isomorphic to the surface of a four dimensional sphere without one point endowed with the natural uniformity. In fig. 1 it can be seen (for a two dimensional X) how the points of the surface of the sphere are to be mapped on the points of X.

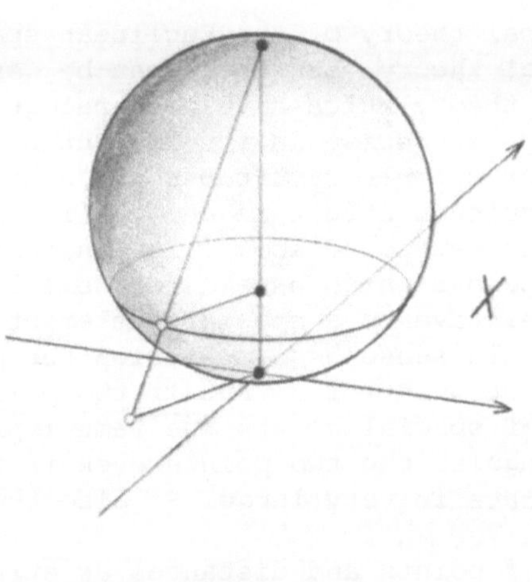

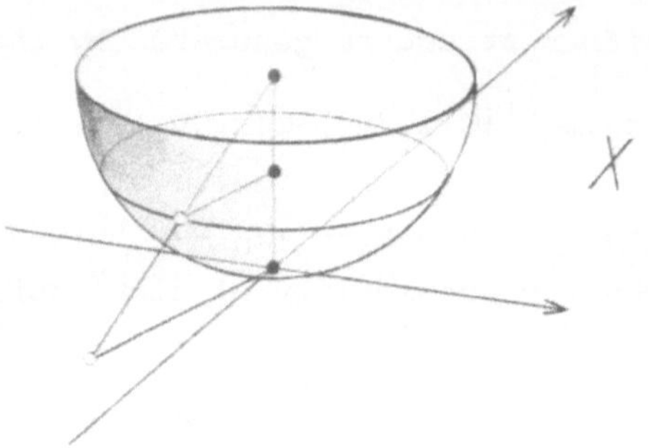

Fig. 2. Completion of $(X, N^{(2)})$

X with $N^{(2)}$ is uniformly isomorphic to one half of the surface of a four dimensional sphere. In fig. 2 it can be seen (for a two dimensional X) how the points of the hemisphere are to be mapped on the points of X.

The completions of X with $N^{(1)}$ or $N^{(2)}$ are different. The completion of X with $N^{(1)}$ is isomorphic to the total surface of the sphere. The completion of X with $N^{(2)}$ is isomorphic to the closed half of the surface.

The reader will find it very instructive to reflect on some special entourages (as imprecision sets) occuring in $N^{(1)}$ and $N^{(2)}$, especially for points x, which have very great distances from a special point $x_o$.

We shall now illustrate how to use $N^{(2)}$ for the two purposes, the imprecision which occur when comparing the mathematical picture with reality, and the imprecision of measurements. We begin with the first.

The mathematical theory of the Euclidean space X is itself a precise mathematical theory. But there can be (and really must be) structures in this theory which have no analogue in reality. The mathematical theory is precise in itself, but not a precise picture of reality. For example structures which have no analogue in reality are the continuum structure and the infinity structure of X as we have already mentioned above. The uniform structure $N^{(2)}$ tells us that two points which have a very small distance cannot be distinguished relative to a specified element $U$ of $N^{(2)}$ as imprecision set. In this sense $N^{(2)}$ qualifies the physical interpretation of the continuity of X. But if two points are very far from a finite set of special points the same imprecision set does not allow to distinguish the two points even if the distance between these two points is very large. So also the infinity of X is qualified by $N^{(2)}$.

The concepts of points and distances as structures of X remain precise as in any mathematical theory. But these concepts cannot describe reality precisely. The elements of $N^{(2)}$ can be used to bridge the gap between the precise concepts and their imprecise application to reality. The concept of the uniformity $N^{(2)}$ is itself a precise concept.

Only the choice of the elements of $N^{(2)}$ used to characterize the imprecision between the mathematical picture and reality is not determined by the theory.

We shall presently illustrate the description of imprecise measurements of distances. We first examine some properties of X endowed with $N^{(2)}$, properties which will be seen to be general properties for all uniform structures introduced in physical theories to describe imprecisions.

$N^{(2)}$ introduced above for X is a metrizable uniform structure. X endowed with $N^{(2)}$ is a precomcapt and separable space.

These three properties seem to me to be fundamental for the following reason:

It is fundamental for physics that every theory can only be tested by finitely many experimental results. I shall call this fact the finiteness of physics. Because of this finiteness every infinity in a mathematical theory has to be compensated by a uniform structure $N$ such that for every imprecision set (as an element of $N$) only finitely many experimental results are sufficient to test the theory. What does this mean?

If M is a set the elements which are representatives of real situations and if $N$ is the uniform structure of M describing imprecision, then:

α) for every $U \in N$ there exist finitely many elements of M, from which all other elements of M cannot be distinguished with the imprecision $U$; i.e. for every $U \in N$ there exists finitely many $x_1, \ldots, x_n \in M$ such that: if $x \in M$ then $(x, x_k) \in U$ for a suitable $x_k$.

This condition α) is equivalent to the statement: M endowed with $N$ is precompact.

β) There exists a denumerable basis of $N$; i.e. the refinement of imprecision sets can be done "step by step".

This condition β) is equivalent to: $N$ is metrizable. α) and β) imply: X is separable.

To illustrate the physical significance of the uniform structures introduced to describe physical imprecisions we will consider for our example of the three dimensional space X the set $P_o(X)$ i.e. the set of subsets of X without the empty set. The elements of $P_o(X)$ can be used as representatives of regions in the real physical space; for instance an element of $P_o(X)$ can be used as a representative of the region which a special body fills; another element of $P_o(X)$ can be used as representative of the trace of the motion of a "small" body, for instance a Kepler ellipse represents the trace of the trajectory of a planet. Since we can distinguish points in X only by elements of $N^{(2)}$ defined above (the idealized form of $N^{(2)}$ allows in the limit to distinguish all points of X) it is plausible to use as uniform structure for $P_o(X)$ the following one: $\tilde{N}^{(2)}$ as generated by the sets:

$$\tilde{U} = \{(A,B) \mid A,B \in P_o(X), A \subset U(B), B \subset U(A), U \in N^{(2)}\}.$$

For $U \in N^{(2)}$ and $B \in P_o(X)$ the set $U(B)$ is defined by

$U(B) = \{x \mid \text{there is an } x' \in B \text{ such that } (x,x') \in U\}$.

$\tilde{N}^{(2)}$ does not separate $P_o(X)$! This means there are two sets A,B which can't be separated by any of the imprecision set of $N^{(2)}$. Also the idealization of smaller and smaller sets of $N^{(2)}$ does not enable one to distinguish physically all subsets of X. This shows that the subsets of X cannot be physically real in every respect.

The uniform structure $\tilde{N}^{(2)}$ implies that we have to divide $P_o(X)$ into equivalence classes. An equivalence class is here the set of all those elements of $P_o(X)$ which cannot be separated by $\tilde{N}^{(2)}$. If B is a subset of X and $\bar{B}$ the closure of B in X, then B and $\bar{B}$ are in the same class. Two closed and different subsets $B_1$ and $B_2$ of X are separated by $\tilde{N}^{(2)}$. So every class can be represented by the unique closed subset B which is element of the class.

This example of $P_o(X)$ endowed with $\tilde{N}^{(2)}$ demonstrates how physically unrealistic mathematical structures can be eliminated.

## III. IMPRECISION OF MEASUREMENT

A uniform structure is not only suitable to describe the imprecision between the mathematical picture and reality, but also to describe imprecisions of measurements. And we want to make the assumption that the same uniform structures can be used to describe the imprecisions of measurements.

But what is the source of imprecisions of measurements?

There is much confusion about the concept of imprecision of measurement. The notion that there are precise quantities in nature but that we cannot avoid making errors in measuring them is in my opinion a fiction. This fiction must be replaced by a theory of the measuring apparatus.

Only such a theory of the measuring apparatus can explain what we mean by an imprecision of measurement. Such theories show that there are two sources producing imprecisions of measurements.

The first source is the imprecision between the mathematical picture and reality discussed above. There are such imprecisions also in the theory of measuring methods. The second source is the finding that measuring apparatuses (in most cases) do not measure what one "wanted" to measure. It is in general difficult enough to construct apparatuses which measure something which is not far away from the quantity one has in mind. These last "imprecisions" are consequently differences between the real measurement and the desired measurement. This last fact plays an important role in quantum mechanics if one "tries" to measure together quantities which can't be measured precisely together.

It is our opinion that all these imprecisions can be described by uniform structures in the theory without going back in all details to the complicated theories of measurement. It is only necessary to give a value for the so-called "error" of measurement, a value gained from the theories of measurement.

An example will illustrate the use of uniform structures of imprecision to describe an imprecision of measurement. We use again the same space X with the distance function d. We had endowed X with the uniformity $N^{(2)}$. We now endow $\mathbb{R}_{0+}$ (as the picture of the measured values of d) with the uniformity generated by the metric

$\Delta(\alpha_1,\alpha_2) = \arc\tan \frac{|\alpha_1-\alpha_2|}{D}$. For the product set $X \times X \times \mathbb{R}_{0+}$ we then choose the product uniformity. For every element W of this product uniformity we define the following "smeared out" relation as a subset of $X \times X \times \mathbb{R}_{0+}$

$$\delta_W = \{(x_1,x_2,\alpha) \mid \text{there are elements } x_1',x_2',\alpha' \text{ such that } ((x_1,x_2,\alpha),(x_1',x_2',\alpha')) \in W \text{ and } d(x_1',x_2') = \alpha'\}.$$

If we make a measurement of the distance between two spots in real space (represented by $x_1,x_2$) then the set W is determined on the one hand by the imprecisions in fixing the spots, and on the other hand by the imprecision of the measured value $\alpha$, an imprecision which is defined by the theory of the specific measuring method. To compare the theory with the experimental result of the measurement we shall not write down this result $\alpha$ of the measured distance between the two spots $x_1,x_2$ in the form

$$d(x_1,x_2) = \alpha$$

but instead will write down the "smeared out" relation:

$$(x_1,x_2,\alpha) \in \delta_W.$$

It is sometimes imagined that the concept of distance in the mathematical theory is smeared out. This is not the case. Only in comparing theory and reality (i.e. experiments) do we use a specific smeared out relation $\delta_W$ where W is determined by the specific experiments.

It may happen that we get no contradictions between theory and experiment if we use the imprecision sets W determined by the measuring methods alone. We say then that the theory is "good enough" relative to the precision of measurements achieved until now.

But if the measuring methods are refined it may happen that the finer imprecision sets W lead to contradictions. Then it is necessary to chose imprecision sets greater than those which result from the theory of the measuring processes alone. Such a situation will be a motivation to improve the theory; that is, to adopt a more comprehensive theory. It is a mistake to take this situation of contradictions between refined measurements and theory as the only motivation to look for new theories.

If we have achieved a more comprehensive theory then we can recover those imprecision sets of the older theory which are determined by the imprecision between its mathematical picture and reality. One can determine these imprecision sets for the older theory because the more comprehensive theory will give a more precise picture of reality than the older one. We have not the time to go into details of the problem of "imprecise" relations between two theories. The reader interested in this feature is referred to [1] § 8.

Our reflections on imprecision of measurements implied that a measurement is carried out only one time. The imprecision of measurement is then determined by the theory of the measuring method which is used. Indeed, many measurements cannot be repeated. For instance we measure the position of a bullet at a certain time using flashlight and a camera the measurement can't be repeated since the bullet has moved. If we measure microsystems no measurement of the same system can be repeated since the interaction with the measuring system drastically affects the microsystem. Each measurement of a microsystem is similar to a measurement of a bullet where the bullet is shot in a sandbag.

It cannot be denied that sometimes one tries to repeat macroscopic measurements with the intention of improving their precision. Such repetitions are possible only if the measured quantity does not change (even under the influence of the measurement), or if the quantity to be measured can be reproduced with a much higher precision than the precision of measurement itself. This last situation arises if the precision of preparing the same situation is much higher than the precision of measurement. Here we encounter a new concept of precision resp. imprecision, namely that of precision of preparation. We will come back to this problem a little later.

If one wants to account for the improvement in precision by repeated measurement it is necessary to have a statistical theory of the measuring apparatus. We will illustrate this for the example of a measurement of distances. $m(x_1,x_2;A)$ may be a measure (depending on $x_1,x_2$) on an algebra $\mathcal{A}$ of subsets of $\mathbb{R}_{0+}$ with $m(x_1,x_2;A) \geq 0$, $m(x_1,x_2;\mathbb{R}_{0+}) \leq 1$, $m(x_1,x_2;A)$ is the probability that the registered value $\alpha$ lies in A. Since the uniform structure of $X \times X \times \mathbb{R}_{0+}$ introduced above describes the imprecision of measurement, it is necessary that $m(x_1,x_2;A)$ is given by a density $\rho(x_1,x_2;\alpha)$ of the form

$$m(x_1,x_2;A) = \int_A \rho(x_1,x_2;\alpha)\,d\alpha$$

and that $\rho(x_1,x_2;\alpha)$ is a uniformly continuous function.

We shall not discuss the use of repeated measurements, together with the knowledge of $\rho(x_1,x_2;\alpha)$ to improve the precision of measurements. There are many books in which these problems are treated.

With these considerations we conclude the treatment of the two problems: imprecisions between theory and reality; imprecisions of measurement. We will go over to the other two problems characterized by such concepts as the Heisenberg uncertainty principle and incommensurability.

## IV. THE HEISENBERG UNCERTAINTY PRINCIPLE AND INCOMMENSURABILITY

The Heisenberg uncertainty relation is a special case of imprecisions of preparations relative to registrations. There is no method of preparing microsystems which yields sharp values for all observables (decision observables; see [2] IV). We can say that every preparing method must be imprecise relative to some registration of decision effects ([2] IV, § 8.3).

But these imprecisions of preparation relative to registration have nothing to do with the imprecisions of measurement or imprecisions between theory and reality treated above. On the contrary the Heisenberg uncertainty relation $\Delta p\ \Delta q \geq \frac{\hbar}{2}$, valid for the dispersions $\Delta p$ and $\Delta q$ in any prepared ensemble, is deduced in the framework of the theory under the assumption that the observables position and momentum are measured with absolute precision. If the real measurements of position and momentum are imprecise, then the imprecisions of the measurements will add to the dispersions $\Delta p$ and $\Delta q$. I know that it is possible to get the impression - reading some books - that the Heisenberg uncertainty relations has something to do with imprecise measurements. But this is not the case.

Moreover, in preparing macrosystems we cannot avoid finite imprecisions which cannot be reduced in size. This is often forgotten. Macrophysics is essentially based on these imprecisions of preparation which in principle cannot be avoided (see [3] XV, [4], [5] X and [6]). The popular conception that macrophysics results from quantum mechanics in the limit $\hbar \to 0$ can only be upheld in the following sense: All possible preparations of macrosystems are much more imprecise than the Heisenberg uncertainty relation postulates. The preparations of macrosystems are so imprecise that the Planck constant $\hbar$ can be neglected.

As we have seen an imprecision of preparation has nothing to do with imprecisions between theory and experiment. These imprecisions of preparation say something about the real possibilities to prepare microsystems. The other quantum mechanical statement, that position and momentum can be measured together only with finite precision, has indeed to do with imprecisions of measurement, especially with imprecisions arising from the fact that the measuring apparatus generally does not measure the desired quantity. To measure precisely and together position and momentum is not possible ([2] IV, § 3). However it is possible to measure observables together which approximate position and momentum.

Since the measuring process for microsystems can only be described with the help of quantum mechanics, the observables position and momentum must be given a quantum mechanical definition (see e.g. [2], § 4). Therefore it is possible also in quantum mechanics to say what an imprecise measurement is and what measuring-together means. From the considerations in [2] IV, § 3 it follows that measuring-together is defined by a Boolean algebra $\sum$ and a measure $\sum \xrightarrow{F} [0,1]$ where $[0,1]$ is the set of all self adjoint operators A with $0 \leq A \leq 1$. As an imprecise measuring-together of position and momentum we shall define a measure $\sum \xrightarrow{F} [0,1]$ where $\sum$ is the Boolean algebra of the Borel sets of the 6-dimensional phase space of one particle. In addition we postulate that the transformations of the measure are canonically connected with the transformations of the subsets of the phase space by Galilean transformations. It is possible to characterize the class of all such measures (see [7]). We do not go into details here, since we are only interested in the meaning of imprecision in this context.

Let $\sigma$ be an open subset of the phase space, v be the projection of $\sigma$ into the real space X and $\pi$ be the projection of $\sigma$ into the momentum space. The well known position operator defines a projection operator E(v) which represents the (decision) effect that the position of the particle is in v. A projection operator $P(\pi)$ is similarly defined, which represents the (decision) effect that the momentum of the particle is in $\pi$.

If there is an imprecision set $\mathcal{U}_1$ ($\mathcal{U}_1 \in \mathcal{N}^{(2)}$) (independent of $\sigma$) for X and similarly an imprecision set $\mathcal{U}_2$ for the momentum space and a small number $\varepsilon > 0$ ($\varepsilon \ll 1$) such that

$$F \leq E(\mathcal{U}_1 v) + \varepsilon 1$$

$$F \leq E(\mathcal{U}_2 \pi) + \varepsilon 1$$

then we say that $\sum \xrightarrow{F} [0,1]$ is an imprecise measuring-together of position and momentum with the imprecision $\mathcal{U}_1$ of position and $\mathcal{U}_2$ of momentum, and with $\varepsilon$ as an upper bound of the probability of finding deviations bigger than these imprecisions $\mathcal{U}_1$, $\mathcal{U}_2$.

In this article we could only give a few examples to demonstrate that uniform structures of physical imprecision are essential in physical theories if we want to discuss the problems of interpretation. It must be stressed again, that the uniform structures do not make the mathematical concepts of the theory imprecise. Rather uniform structures are mnemonic devices which help us avoid confusing mathematical idealizations with physical reality.

REFERENCES


1 G. Ludwig, Die Grundstrukturen einer physikalischen Theorie, (Springer, 1978)

2 G. Ludwig, Foundation of Quantum Mechanics (2 volumes; Translation in English in preparation; Springer)

3 G. Ludwig, Einführung in die Grundlagen der theoretischen Physik (4 volumes; Vieweg, 1974-1979)

4 G. Ludwig, The Connection Between the Objective Description of Macrosystems and Quantum Mechanics of "Many Particles", in Essays in Honor of Wolfgang Yourgrau. Alwyn van der Merve, editor (Plenum Press, New York, 1981), to be published.

5 G. Ludwig, An Axiomatic Basis of Quantum Mechanics (in preparation; Springer)

6 R. Werner, Article in this volume

7 R. Werner, Covariant observables on phase space and continuity properties of operators under Galilei-transformations. (in preparation)

# CAUSALITY IN STATISTICAL THEORIES AS AN EXAMPLE FOR THE IDEALIZATION OF PHYSICAL LAWS

H. Neumann

Fachbereich Physik, Philipps-Universität Marburg
Renthof 7
3550 Marburg
Federal Republic of Germany

When two physical theories describe the physical "facts" equally well, Poincaré has proposed to always adopt the "simpler" theory. While there may be doubts about the applicability of this proposal as a general criterion, it often describes the behaviour of scientists even in the case when the two theories do not describe the facts "equally well": In practical work physicists usually tolerate a larger margin of inaccuracy if they can only use a "simple" theory. Simplicity may concern the mathematical part of the theory as well as its physical content. The precise meaning of "simplicity" and "approximation" is generally quite clear for theories which are "simplifications" of more a comprehensive theory. But in most cases no such theory is available as a standard of comparison. Yet even in the framework of one given theory physicists often introduce "simplifying assumptions", called "idealizations" in the work of Ludwig. The purpose of this paper is to discuss some problems related to idealizations and the role of physical imprecision and approximation in this context.

Since the modification of a principal assumption or axiom of a theory implies the transition to another theory, idealization is strictly speaking a problem of intertheoretical approximation. But in view of the above remarks we shall not attempt to give an adequate definition of the relation "physical theory $\mathcal{PT}_1$ is an idealization of $\mathcal{PT}_2$". Instead, we shall restrict our considerations to the special case: "an axiom $\mathcal{P}_1$ is an idealization of $\mathcal{P}_2$", this relation will be illustrated by an example which is itself of physical interest.

In the study of idealization, an examination of the mathematical description of experimental tests of a physical theory turns out to be necessary. In particular, the application of the definitions given below requires the proof of a metatheorem concerning the general structure of observational reports $\mathcal{A}$ (Abbildungsaxiome). However, serious questions concerning this structure will remain unsolved.

We try to give a precise definition of $P_1$ being an idealization of $P_2$ in the framework of the description of physical theories by Ludwig [2]. Let $MT_\Sigma$ be a mathematical theory of the species of structures $\Sigma$ and let $MT_\Sigma$ be an axiomatic basis for a physical theory. We consider two additional axioms $P_1$ and $P_2$ in $MT_\Sigma$. $P_1$ and $P_2$ are assumed to admit a physical interpretation. In $MT_\Sigma$ a uniform structure N is defined which describes the physical imprecision. In a physical test of $MT_\Sigma$ a set of axioms $A_{\mathcal{U}}$ is added, where the interpreted relations are smeared out with an imprecision set $\mathcal{U} \in N$. In the following definition $A_{\mathcal{U}_1}$ and $A_{\mathcal{U}_2}$ denote the same set of axioms blurred with different imprecision sets $\mathcal{U}_1$ and $\mathcal{U}_2$, respectively. If $\mathcal{U}_1 \subset \mathcal{U}_2$, $A_{\mathcal{U}_1} \Rightarrow A_{\mathcal{U}_2}$.

Definition: $P_1$ is called a <u>pure idealization</u> of $P_2$ in $MT_\Sigma$ if $P_1 \Rightarrow P_2$ and for all $\mathcal{U} \in N$ there is a $\mathcal{V} \in N$ with $\mathcal{V} \subset \mathcal{U}$ such that for all A a contradiction of $P_1$ in $MT_\Sigma A_{\mathcal{U}}$ implies a contradiction of $P_2$ in $MT_\Sigma A_{\mathcal{V}}$. (i.e. from [$\neg P_1$ is a theorem in $MT_\Sigma A_{\mathcal{U}}$] follows [$\neg P_2$ is a theorem in $MT_\Sigma A_{\mathcal{V}}$]).

Though this definition might be sufficiently motivated by physical intuition, a concrete test of this definition in general requires complicated investigation of $MT_\Sigma$ and the structure of the axioms A. In the axiomatic foundation of quantum mechanics by Ludwig (see for example [3]) axiom AVid seems to be a pure idealization (of an "axiom" which is a theorem).

We also want to consider the case that the axiom $P_1$ is physically indistinguishable from $P_2$ only in tests with imprecision sets $\mathcal{U} \supset \mathcal{U}_o$ for a certain $\mathcal{U}_o \in N$. For example this can occur if $P_2$ depends on $\mathcal{U}_o$. If a finite imprecision set is to be included in a physical theory there is a need for a more comprehensive theory which gives an explanation for $\mathcal{U}_o$.

Definition: $P_1$ is called <u>physically indistinguishable</u> from $P_2$ in $MT_\Sigma$ with respect to $\mathcal{U}_o \in N$ if there is a $\mathcal{V} \in N$ with $\mathcal{V} \subset \mathcal{U}_o$ such that for all A a contradiction of $P_1$ in $MT_\Sigma A_{\mathcal{U}}$ implies a contradiction of $P_2$ in $MT_\Sigma A_{\mathcal{V}}$ and the same holds if $P_1$ and $P_2$ are interchanged.

If $P_1 \Rightarrow P_2$, we call $P_1$ an idealization of $P_2$ in $MT_\Sigma$ with respect to $\mathcal{U}_o$.

As an example, we consider causality axioms in the statistical description of the interaction between preparing and recording apparatuses. Mathematical details and other aspects of this example may be found in [4].

We start from a general statistical description of experiments by ensembles and effects. Every individual experiment may be split into preparation and registration. We may confine ourselves to considering only registrations with two possible outcomes 0 and 1.

The preparations giving the same statistics with respect to all of the considered recording methods are collected into equivalence classes called ensembles. The corresponding equivalence classes of recording methods are called effects. Let $\mu(W,F)$ denote the probability for measuring the outcome 1 of the effect F in the ensemble W. (comp. [5]).

The set K of ensembles can be considered as a base of a base-normed ordered Banach space B and the set L of effects can be considered as the order interval [0,1] of the dual B' such that $\mu(W,F)$ coincides with the canonical bilinear form $\langle W,F\rangle$ on $K \times L$.

We include the space-time structure into this setting by a mathematical description of the spatio-temporal transformations of the preparing and recording procedures relatively to each other. In addition we assume that the macroscopic processes of preparing and recording occur in bounded regions of space-time. According to the principle of causality in special relativity, the preparing apparatus cannot influence the recording apparatus if the corresponding regions of space-time are space-like separated. In this case the registration will yield the same frequency as if there were no preparing apparatus, i.e. the outcome "1" occurs with "vacuum"probability.The preparation of "vacuum" is defined as the protection of the recording apparatus from uncontrolled exterior influences.

The mathematical formulation of these ideas is summarized in

Axiom C:

(1) There is a weakly continuous representation $\alpha$ of the restricted Poincaré group $P_+^\uparrow$ by linear transformations of B with $\alpha(K) \subset K$. Let $\bar{\alpha}$ denote the dual representation in B' defined by $\langle W,\bar{\alpha}_g F\rangle := \langle \alpha_{g^{-1}} W,F\rangle$.

(2) For all bounded open sets $\sigma \subset \mathbb{R}^4$ there are $K(\sigma) \subset K$ and $L(\sigma) \subset L$ such that $K_f = \bigcup_{\sigma\in\mathcal{O}} K(\sigma)$ separates on L and $L_f = \bigcup_{\sigma\in\mathcal{O}} L(\sigma)$ separates on K. ($\mathcal{O}$ denotes the set of bounded open subsets of $\mathbb{R}^4$.)

(3) For all $g \in P_+^\uparrow$, we have $\alpha_g K(\sigma) \subset K(g\sigma)$, $\bar{\alpha}g\, L(\sigma) \subset L(g\sigma)$.

(4) There is an ensemble $W_o \in K$ (vacuum) such that for $W \in K(\sigma)$ $F \in L(\sigma')$ and $\sigma$ spacelike to $\sigma'$ we have

$\alpha$) $\langle W,F\rangle = \langle W_o,F\rangle$.

Let us briefly comment on (2): $K(\sigma)$ is interpreted as the set of ensembles which can be prepared within the space-time region $\sigma$. $L(\sigma)$ is interpreted similarly. The separation properties express that approximately all ensembles can be prepared, and approximately all effects can be recorded in bounded regions.

Since in the quantum case to be discussed later (4α) has some surprising consequences one might consider less idealized causality postulates. An obvious weakening taking into account the imprecision of probability measurements would be the replacement of (4α) by

$$(4\beta) \quad |\langle W,F\rangle - \langle W_o,F\rangle| < \varepsilon_o$$

for a finite but sufficiently small $\varepsilon_o > 0$.

Let us also consider

$$(4\gamma) \quad |\langle W,F\rangle - \langle W_o,F\rangle| < \varepsilon(d(\sigma,\sigma'))$$

where ε is a function $\varepsilon: \mathbb{R} \to \mathbb{R}_+$ and

$$d(\sigma,\sigma') = \inf_{x\in\sigma, x\in\sigma'} \{-(x_o-x'_o)^2 + (\vec{x}-\vec{x}')^2\}.$$

For an experimental test with a certain fixed imprecision set $\mathcal{U}_o$ postulates (4α), (4β) and (4γ) with appropriately chosen $\varepsilon_o$ and $\varepsilon(d)$ should be physically equivalent with respect to $\mathcal{U}_o$. (4α) should be an idealization of (4β) and (4γ).

By means of (4γ) an imprecision for the relation "σ spacelike to σ'" can be taken into account. For $d \to 0$ one might assume $\varepsilon(d) \to 1$ and for $d \to \infty$ one might assume $\varepsilon(d) \to 0$. For example $\varepsilon(d) = \exp(-md)$ for a suitable constant $m > 0$.

To give a motivation for considering (4β) or preferably (4γ) instead of (4α) we shall briefly discuss the quantum case of the sketched statistical description. If the carriers of the interaction between the preparing and the recording apparatuses are quantum systems, we require

Axiom Q: $K = \{W \in B_h(H) / W \geq 0,\ \mathrm{tr}\, W = 1\}$

$$L = \{F \in B_h(H) /\ 0 \leq F \leq 1\}$$

where $B_h(H)$ is the class of Hermitian bounded operators on a Hilbert space H. The canonical bilinear form is given by $\langle W,F\rangle = \mathrm{tr}\,(W\cdot F)$.

As a consequence of Axiom Q there is a unitary representation $\mathcal{U}$ of the Poincaré group $P_+^\uparrow$ in H such that

$$\alpha_g(W) = \mathcal{U}_g W \mathcal{U}_g^*, \quad \overline{\alpha}_g(F) = \mathcal{U}_g F \mathcal{U}_g^* \text{ for } g \in P_+^\uparrow.$$

The subgroup of space-time translations determines uniquely a projection valued measure E on $\mathbb{R}^4$ such that for all $x \in \mathbb{R}^4 \subset P_+^\uparrow$

$$\mathcal{U}_x = \int e^{ipx}\, dE(p) \quad \text{with } px = x_0p_0 - \vec{p}\,\vec{x}.$$

The support of E is characterized by the decomposition of the representation of $P_+^\uparrow$ into irreducible parts and the resulting mass distribution. In quantum field theory one usually assumes supp $E \subset \Gamma$, where $\Gamma = \{p \in \mathbb{R}^4 \; / \; p \geq |\vec{p}|\}$ is the forward light cone in $\mathbb{R}^4$. This spectral condition is to exclude that a particle may lose energy indefinitely producing an infinite number of particles or quanta of positive energy.

The spectral condition and causality postulate (4α) exclude the existence of non-trivial counters in bounded space-time regions in the sense of the following definition and theorem.

Definition: An effect $F \in L$ is called a <u>counter</u> if $\langle W,F \rangle \geq \langle W_o,F \rangle$ for all $W \in K$.

Theorem: Assume Axiom C1 through C4α, Axiom Q and the spectral condition. If $F \in L(\sigma)$ for some bounded $\sigma \subset \mathbb{R}^4$ and F is a counter, we have $F = \lambda \cdot 1$ with $\lambda \in \mathbb{R}$.

According to this theorem the idealization of causality postulate (4α) and the idealization of local counters are contradictory. If one does not want to dispense with the existence of local counters, the causality postulate (4α) has to be weakend - preferably according to (4γ).

REFERENCES


[1] G. Ludwig: Imprecision in Physics.
(This volume)

[2] G. Ludwig: <u>Die Grundstrukturen einer physikalischen Theorie</u>, Springer (1978)

G. Ludwig: Axiomatische Basis einer physikalischen Theorie und theoretische Begriffe. To appear in: Zeitschrift für allgemeine Wissenschaftstheorie.

[3] G. Ludwig: An axiomatic basis of quantum mechanics, in H. Neumann (ed.): <u>Interpretations and Foundations of Quantum Theory</u>, BI-Wissenschaftsverlag (1981)

[4] H. Gerstberger, H. Neumann, R. Werner: Makroskopische Kausalität und relativistische Quantenmechanik, in J. Nitsch , J. Pfarr, E.-W. Stachow (eds.), <u>Grundlagenprobleme der modernen Physik</u>, Bibliographisches Institut, Mannheim, 1981.

H. Neumann, R. Werner: Causality between Preparation and Registration Processes in Relativistic Quantum Theory.
(To be published)

[5] R. Werner: Approximate Embeddings in Statistical Mechanics
(This volume)

# APPROXIMATE EMBEDDINGS IN STATISTICAL MECHANICS

R. Werner

Fachbereich Physik

Postfach 44 69, 4500 Osnabrück
Federal Republic of Germany

## I. INTRODUCTION

It is part of the scientific folklore of industrial societies that macroscopic systems like gases and solids, stars, tables and animals are "composed of" atoms and molecules. Most of us are so accustomed to this atomistic view that we tend to overlook the conceptual problems inherent in the expression "composed of". On second thought we are faced with two incommensurable views of the same individual systems: For example, when we use the microscopic description of a gas in terms of many atoms we are unable to say what the temperature of the gas at a given spatial point is. (Most of the time there won't even be a single atom near that point.) On the other hand, when using a macroscopic (thermodynamic) description it is impossible to say where all the particles are. Yet there must be an intimate connection between the two descriptions: For example we are able to identify the substance "water" either by its macroscopic properties (density, viscosity, freezing and melting points etc.) or by the structure of its molecules ($H_2O$).

The physical theory concerned with these interrelations is statistical mechanics. In this paper I will study some fundamental problems of this discipline. Since statistical mechanics by its very definition is the study of the relation between microscopic and macroscopic theories it will be useful to review some basic structures of the two types of theories involved.

Thermodynamics (or rather: "thermodynamical theories" since there is no accepted general framework for non-equilibrium thermodynamics comprising all theories describing special types of material) is a "classical" theory in the following sense: At a given time each individual system may be described completely by the specification of all its properties; that is by its "state parameters" such as temperature, pressure and densities at all points in the system. The set of possible values of these parameters is called the phase space (or state space) $\Omega$ . Thus each individual system is described by a point in phase space. One now tries to derive laws of motion. In same cases it is sufficient to know the state of the system at one time to predict its state at any later time. Then the dynamical laws will be called deterministic. Sometimes, as in the theory of Brownian motion or other macroscopically visible fluctuations, one can predict the future state of the system only with a certain probability. Since this case is clearly more general I shall assume in the sequel that thermodynamics is a statistical theory in which the dynamics is given by an evolution equation for probability densities over the phase space $\Omega$. Typically the time evolution is irreversible, i.e. for large times systems approach an equilibrium state.

The microscopic theory we are going to use will have to describe adequately single molecules and collections of them. Its basic structure therefore must be that of quantum mechanics and is thus non-classical. Sometimes one also uses a many particle version of classical Hamiltonian mechanics as a microscopic theory. But in principle this should be regarded either as a means for simplifying quantum mechanical computations or as a means of illustrating some general features of many-particle descriptions. The dynamics of a quantum mechanical theory is generated by a Hamiltonian operator in a Hilbertspace $\mathcal{H}$. It takes into account the numbers and types of molecules and the interactions between them. This time-evolution is reversible in the sense that for every physically possible process the time-reversed process is also admitted by the theory. For systems contained in some finite volume (i.e. for all systems we shall ever want to describe) time-evolution is also recurrent. This means that every initial state recurs to an arbitrary precision after some sufficiently large time.

Any approach to the foundations of statistical mechanics will eventually have to explain these two structural differences between macroscopic and microscopic description: How can macroscopic systems be described "objectively" in a classical theory in spite of the quantum mechanical nature of the microscopic description? And why do macroscipic systems exhibit an irreversible approach to equilibrium whereas the microscropic dynamics is reversible and recurrent? In this paper I shall not attempt to "explain" these macroscopic structures. But I shall present an axiomatic formulation of statistical mechanics (due to Ludwig [1]) and show that the structural differences described here do not lead to a contradiction in this approach. In the discussion the concept of intertheoretic approximation will play a decisive role.

## II. THE CONCEPT OF EMBEDDINGS

To study the relations between the two descriptions, we have to formulate both in a common framework. This framework should not be a mere mathematical scheme but should also include a physical interpretation of the theories. Since both theories are statistical, I shall use Ludwig's concept of "statistical theories" [2] as a tertium comparationis.

The starting point of this concept is the following idea: Each individual experiment may be split into two parts. First, systems are prepared according to some laboratory procedure. This procedure may involve technical "production" processes as well as the selection of objects from a larger class of natural or artificial objects according to specified criteria. In any case the procedure is assumed to be given as a list of detailed prescriptions in a laboratory manual. After the preparation, a measurement is performed according to another procedure. Included in the description of this procedure is a statement as to when the experiment is to be considered as terminated, and which of the possible outcomes has then occurred. For our purposes it is sufficient to consider registration procedures with only two possible outcomes, say "+" and "-". If the experiment is repeated many times according to the same pair of procedures (preparation and registration) the result "+" will occur with a certain relative frequency. These frequencies (or probabilities) are the basic quantities of the theory. They must be assumed to be reproducible. (For this, the procedures must be sufficiently well specified, since preparing procedures such as "pick up any stone" can hardly lead to reproducible rates for temperature measurements.)

Different preparing procedures which generate the same frequencies for all available registrations will be called equivalent. An equivalence class of preparing procedures will be called an ensemble or (statistical) state (not to be confused with a point in phase space, i.e. the "state" of a classical system). The set of states will be denoted by K and carries a natural convex structure due to the possibility of "mixing" procedures. Analogously the set of equivalence classes of registration procedures (called "effects") becomes a convex set L. The probabilities are then given by a biaffine function $\langle\cdot,\cdot\rangle : K \times L \to [0,1] \subset \mathbb{R}$. It is very important to note that the definition of statistical states depends on the class of available registration procedures. When a new method of registration is invented and included into the set of admissible procedures the appearance of the theory in terms of states and effects may change drastically. (An analogous remark pertains to the definition of effects.) Leaving the system to itself for some specified time before applying a certain measurement defines a transformation in the set of registration procedures. On the level of effects this induces a semigroup of operators $V_t : L \to L$ $(t \geq 0)$ called the time evolution.

In this framework the structures described in the introduction may be rephrased mathematically as follows:

For the macroscopic theory $K_M := \{\rho \in L^1(\Omega,\mu) \mid \rho \geq 0; \int \rho d\mu = 1\}$ is the set of probability densities over a phase space $\Omega$. It follows that the set of effects may be represented as $L_M := \{f \in L^\infty(\Omega,\mu) \mid 0 \leq f \leq 1\}$ and the probability function as $\langle\rho,f\rangle_M := \int \mu(d\omega)\rho(\omega)f(\omega)$. The time evolution will be denoted $V_t$. Irreversibility means that the semigroup $V_t$ cannot be extended to a group defined for all times, including negative times. Approach to equilibrium means that the limit $\lim_{t\to+\infty} V_t$ exists.

For the microscopic theory $K := \{W \in B(H) \mid W \geq 0; \mathrm{tr}W = 1\}$ for some Hilbertspace $H$. Then $L := \{F \in B(H) \mid 0 \leq F \leq 1\}$ and the frequency function is given by $\langle W,F\rangle := \mathrm{tr}(W \cdot F)$. The time evolution $U_t$ is generated by the Hamiltonian H as $U_t F := e^{-iHt} F e^{iHt}$. The recurrence property is equivalent to H having a discrete spectrum.

In concrete applications these general assumptions are supplemented by detailed information about the structure of the Hamiltonian, the space $\Omega$ as a set of functions depending on position and the macroscopic time evolution. But in this article none of these refinements will be considered.

The main advantage of the formulation presented so far is that both theories are build up from directly interpretable elements of the same type (measuring and preparing procedures). Once the relationship between the procedures belonging to the two theories is established the relationship between all other elements is determined by retracing their mathematical construction from these basic elements. But procedures in the sense of laboratory manuals may be specified without any previous knowledge of the theory at hand. (Their description belongs to a pre-theory in the sense of Ludwig). So two theories about the same class of systems can only differ in the set of procedures admitted in their construction.

From what is known about the structure of macroscopic and microscopic theories and their states and effects, we can now get a better idea about the type of measuring and preparing procedures necessary to generate these theories.

Typical experiments performed with macroscopic systems certainly may be accounted for by a macroscopic theory. This is not surprising since macroscopic theories are designed to describe such comparatively simple experiments. For example it is obvious what kind of preparing procedure is needed to produce systems with a given temperature. Other macroscopic states may require the preparation of a complicated distribution of velocities in a liquid. Despite the difficulty of such preparations, their possibility appears as an extrapolation of familiar laboratory techniques.

On the other hand the preparation of a typical microscopic state represented by a density matrix $W \in K$ may require the simultaneous specification of the positions of all $10^{23}$ particles in gas. Even if only a modest accuracy of position (relative to atom-dimensions) is required the list of specifications alone fills any library. To build the corresponding apparatus, let alone to check by statistical experiments whether the aim has been achieved, clearly exceeds human powers. Whether one sees a transition from quantity to quality in this scale of increasing technical difficulties and declares such state preparations impossible, or whether one regards them as possible in principle but difficult in practice, may be a matter of taste. In either case it is clear that none of our experience with macroscopic bodies is based on such preparations and observations, but only on the small subclass of relatively crude procedures that is also used to construct macroscopic theories. Thus the connection between macroscopic and microscopic descriptions is to be developed from the idea that macroscopic procedures may also be considered as microscopic procedures but not conversely. Accordingly, the equivalence classes of preparation procedures induced by all (including microscopic) measuring procedures are strictly smaller than those induced by the macroscopic procedures alone. So we arrive at the following postulate:

There exist linear mappings $S : K_M \to K$ and $T : L_M \to L$ such that $\forall_{\rho\in K_M} \forall_{f\in L_M} \forall_{t\geq 0} \langle\rho, V_t f\rangle_M = \langle S\rho, U_t(Tf)\rangle$.

This equation will be called the embedding equation (EE). It expresses the fact that the probabilities measured in a macroscopic experiment do not depend on whether the procedures used are considered as part of the macroscopic or of the microscopic theory. The inclusion of time translations means that "waiting for a time t before performing a measurement" does not change the macroscopic character of this measurement. The pair (S,T) will be called an embedding. (This terminology agrees with Ludwig's idea of "embedding" as a general intertheoretical relation.)

## III. APPROXIMATE EMBEDDINGS

The first question to be asked now is: Do embeddings exist? Against this, two objections may be raised.

The first objection is classical ("Wiederkehreinwand"): Since the time evolution of a (finite) quantum system is recurrent, the right hand side of (EE) is an almost periodic function of t. Thus (EE) can be satisfied only if for all $\rho \in K_M$, $f \in L_M$ $\langle\rho, V_t f\rangle_M$ is likewise almost periodic. Clearly this contradicts an approach to equilibrium.

But even if (EE) is only postulated for t = 0, i.e. if time evolutions are not considered at all, the equation implies [3] that the phase space $\Omega$ of the macroscopic theory must be discrete (as a measure space). This rules out the existence of continuous state parameters (like temperature) which are however quite typical of macroscopic theories.

Both objections have obvious refutations:
First, the recurrence times are very large. In models they can be seen to be large even compared to the age of the universe. So for practical purposes they should be irrelevant and (EE) may still be satisfied to a very high degree of approximation for a very long time. Secondly, from the point of view of this conference, it is clear that there is no sharp distinction between discrete and continuous theories: Every continuous scale may be approximated arbitrarily well by a sequence of discrete points.

These arguments show that the use of approximations in statistical mechanics is inevitable, so that the equality in the embedding equation has to be replaced by some appropriately chosen concept of approximate equality. If the deviations from equality are very small this clearly makes no difference physically. Hence this replacement does not invalidate the embedding concept itself. In fact the ideas used to motivate this concept already suggest such modifications: Since macroscopic procedures are considered as "relatively crude" microscopic procedures, it seems unnatural to postulate that they lead to the same probabilities with absolute precision.

In statistical mechanics one often uses limiting procedures to demonstrate certain typical aspects of macroscopic systems. For example in the thermodynamic (infinite volume) limit the recurrence times also go to infinity. This result shows that recurrence times are indeed very large for "sufficiently large" quantum systems. (An explicit numerical estimate is of course much harder to give.) Other macroscopic phenomena like phase transitions also become sharply defined mathematically only in this limit. Much insight into the nature of phase transitions has been gained that way, but these results cannot be applied directly to the systems of given finite size on which phase transitions are actually observed. In ther terminology of [4] the statistical mechanics of finite systems require the concept of "proximation" rather than "approximation": The limiting case is relevant only to the degree that it is "close" to the finite situation.

The formulation of results about finite "proximations" clearly requires a more detailed and explicit specification of "measures of proximity" than the analysis of asymptotic approximations. For embeddings I shall describe one such type of measures below. With respect to this definition of proximity one may reconsider the question asked at the beginning of this paragraph. The existence problem of approximate embeddings will then be given as complete an answer as can be expected on this level of generality: The appeal to the approximative nature of embeddings is also sufficient to guarantee their existence and therefore the compatibility of macroscopic and microscopic decriptions.

As a first attempt to define a measure of proximity we consider the function $\Delta_o(t) := \sup_{\rho\in K_M} \sup_{f\in L_M} |\langle\rho, V_t f\rangle_M - \langle S\rho, U_t Tf\rangle|$.

If this function is very small for a long time (compared to the time in which the macroscopic systems practically reach equilibrium) the probabilities on either side of (EE) will be physically indistinguishable. But this requirement is still too stringent. Too see this, suppose that $\rho$ corresponds to a measure on $\Omega$ concentrated near the point $\omega_1 \in \Omega$. Suppose that the state $S\rho$, when tested with the effects $Tf$ (mathematically speaking: the state $T'S\rho$) corresponds to a measure concentrated near another point $\omega_2 \in \Omega$. By choosing a function which is equal to 1 near $\omega_1$ and zero near $\omega_2$ one sees that $\Delta_o(O) = 1$. Hence the embedding $(S,T)$ is qualified as "worst possible" even if the points $\omega_1$ and $\omega_2$ are arbitrarily close to each other and therefore equivalent physically.

To improve this situation we have to take into a account as an additional structure in the space $\Omega$ a uniform structure describing the possibility to discern points of $\Omega$ by physical experiments. As we are interested in numerical estimates, it will be convenient to choose a definite metric d to parametrize the "size" of members of this uniformity. For reasons discussed in [5] we shall assume that the metric space $(\Omega,d)$ is precompact. This means that at every given accuracy we can only distinguish finitely many essentially different points.

One method of distinguishing points of $\Omega$ is the application of the registration procedures introduced above. So the metric d must be consistent with those procedures. This reveals a discrepancy in the argument presented above. If the effect f may be realized by a certain registration procedure, this procedure separates the points $\omega_1$ and $\omega_2$ with probability one. So they cannot be "arbitrarily close" in the metric d. Conversely, when they are very close the function f cannot correspond to an actual measurement but only to some idealized element of L (as for example a weak limit of more realistic elements). We conclude that the typical effects, corresponding to those measurements for which the embedding equation is approximately valid, are slowly varying functions with respect to d.

We define the class of functions not varying faster than d as

$$\Lambda_d := \{f\in C(\Omega) \mid \forall_{\omega_1,\omega_2\in\Omega} \ |f(\omega_1)-f(\omega_2)| \le d(\omega_1,\omega_2)\}.$$

For the special function $\rho$ and the embedding S, T described above we obtain: $\sup_{f\in\Lambda_d} |\langle\rho,f\rangle_M - \langle S\rho,Tf\rangle| \approx d(\omega_1,\omega_2)$. Therefore if we require $\langle\rho,V_t f\rangle_M \approx \langle S\rho,U_t Tf\rangle$ only for the functions $f \in \Lambda_d$ we arrive at a measure of proximity expressed in units of the metric d: Set $\Delta(t) := \sup_{\rho\in K_M} \sup_{f\in\Lambda_d} |\langle\rho,V_t f\rangle_M - \langle S\rho,U_t Tf\rangle|$. Then we will call the pair $(S,T)$ an approximate embedding of $(\Omega,d,V_t)$ into $(\mathcal{H},U_t)$ if

$\Delta(t)$ remains small for a sufficiently long time.

## IV. EXISTENCE THEOREMS

In this paragraph I shall assume that a specific macroscopic theory $(\Omega, d, V_t)$ is given. Is it possible then for any given degree of approximation to construct a Hilbertspace $\mathcal{H}$ with time evolution $U_t$ and an approximate embedding $(S,T)$? The answer will of course depend on additional assumptions imposed on $\mathcal{H}$ and $U_t$. At the level of generality we have adopted here it would not make much sense to postulate any specialized structures. So I shall restrict myself to two assumptions that are general properties of finite quantum systems: Firstly the spectrum of the Hamiltonian H has to be discrete. Secondly we will be dealing only with macroscopic systems of finite total energy. Even though the macroscopic energy is not necessarily equal to the Hamiltonian (but only macroscopically equivalent) it seems reasonable to assume that eigenvalues of H which are much larger than the macroscopic energy do not have any influence on the time evolution of macroscopic expectation values. Using spectral theory these two assumptions imply that $\mathcal{H}$ must be finite dimensional. Conversely $\dim \mathcal{H} < \infty$ trivially implies that H is bounded and has discrete spectrum so we are sure to be dealing with the full recurrence problem. In any case the restriction to finite dimensions only makes our results more general since the infinite dimensional cases will follow as straightforward corollaries.

The dimension of a Hilbert space is the largest number of states that may be distinguished from each other with probability one by suitable experiments. Using the metric d we may define similar measures for the "size" of $\Omega$ that will later be compared with the dimension:

Let $(\Omega, d)$ be a precompact metric space and $\varepsilon > 0$.

Then the "stacking number" $N_-(\varepsilon)$ and the "covering number" $N_+(\varepsilon)$ are defined as: $N_-(\varepsilon) := \max\{N \in \mathbb{N} \mid \underset{\omega_1 \ldots \omega_N \in \Omega}{\exists} \ \underset{i \neq j}{\forall} \ d(\omega_i, \omega_j) > 2\varepsilon\}$

$$N_+(\varepsilon) := \min\{N \in \mathbb{N} \mid \underset{\omega_1 \ldots \omega_2 \in \Omega}{\exists} \ \underset{\omega \in \Omega}{\forall} \ \underset{i}{\exists} \ d(\omega, \omega_i) < \varepsilon\}.$$

These numbers have an obvious geometrical meaning related to the arrangement of balls of radius $\varepsilon$ in $\Omega$. So for $\varepsilon > 0$ $N_-$ is the largest number of states that may be distinguished clearly and $N_+$ the smallest possible number of states sufficient to approximate any other state. One always has $N_-(\varepsilon) \leq N_+(\varepsilon)$. Both numbers are finite, due to precompactness of $\Omega$, and go to infinity as $\varepsilon \to 0$ (unless $\Omega$ is a finite set).

In the first step I shall not consider the approximation of time evolutions so that only $\Delta(0) := \sup_{\rho \in K_M} \sup_{f \in \Lambda_d} |\langle \rho, f \rangle_M - \langle S\rho, Tf \rangle|$ is to be estimated:

THEOREM 1

Let $(\Omega,d)$ be a precompact metric space.
Then (1) For all $\varepsilon > 0$ there exist a Hilbert space $\mathcal{H}$ and an embedding of $\Omega$ to $\mathcal{H}$ such that $\Delta(0) \le \varepsilon$ and $\dim \mathcal{H} = N_+(\varepsilon)$.

(2) For any embedding $(S,T)$ to a Hilbert space $\mathcal{H}$ and $\varepsilon > 0$

$$\Delta(0) \ge \varepsilon\left(1 - \frac{\dim \mathcal{H}}{N_-(\varepsilon)}\right)$$

*Proof*: (1) Pick $N_+(\varepsilon)$ points $\omega_1 \ldots \omega_{N_+}(\varepsilon)$ as in the definition of $N_+(\varepsilon)$. Let $\eta_n$ be the point measure at $\omega_n$ and $\{h_n\}$ a set of functions such that $h_n \ge 0$, $h_n(\omega) = 0$ for $d(\omega,\omega_n) > \varepsilon$ and $\sum_n h_n(\omega) \equiv 1$. (Such functions exist since the $\varepsilon$-balls at $\omega_n$ cover $\Omega$). Let $\{P_n\}$ be an orthogonal family of one-dimensional projections in an $N_+(\varepsilon)$-dimensional Hilbert space $\mathcal{H}$ and set $S(\rho) := \sum_n \langle\rho,h_n\rangle P_n$ and $T(f) := \sum_n \langle\eta_n,f\rangle P_n$. Then for $f \in \Lambda_d$:

$$|\langle\rho,f\rangle - \langle S\rho,Tf\rangle| = |\langle\rho,f\rangle - \sum_{nm} \langle\rho,h_n\rangle\langle\eta_n,f\rangle \operatorname{tr} P_n P_m|$$

$$= |\langle\rho,f\rangle - \sum_n \langle\rho,h_n\rangle f(\omega_n)|$$

$$= |\int\rho(d\omega) \sum_n h_n(\omega)\{f(\omega)-f(\omega_n)\}|$$

$$\le \sum_n \int \rho(d\omega)\, h_n(\omega)\,|f(\omega)-f(\omega_n)| \le \varepsilon \sum_n \langle\rho,h_n\rangle = \varepsilon.$$

(2) Let $\omega_1 \ldots \omega_{N_-(\varepsilon)}$ be as in the definition of $N_-(\varepsilon)$ and let $\eta_n$ be the point measures at $\omega_n$. Let $f_n(\omega) := 1 - \varepsilon^{-1} d(\omega,\omega_n)$ for $d(\omega,\omega_n) \le \varepsilon$ and $f_n(\omega) = 0$ otherwise. Then $\varepsilon f_n \in \Lambda_d$, $\sum_n f_n \le 1$ and $\langle\eta_n,f_n\rangle = 1$. Consequently:

$$\Delta(0) \ge \max_n |\langle\eta_n,\varepsilon f_n\rangle - \langle S(\eta_n),T(\varepsilon f_n)\rangle| = \varepsilon(1-\min_n\langle S(\eta_n),T(f_n)\rangle).$$

On the other hand:

$$\min_n\langle S(\eta_n),T(f_n)\rangle \le \frac{1}{N_-(\varepsilon)} \sum_n \operatorname{tr} S(\eta_n)T(f_n) \le$$

$$\le \frac{1}{N_-(\varepsilon)} \sum_n \operatorname{tr} \frac{S(\eta_n)}{\|S(\eta_n)\|} T(f_n) \le \frac{1}{N_-(\varepsilon)} \sum_n \operatorname{tr} T(f_n) \le \frac{1}{N_-(\varepsilon)} \operatorname{tr} T(1) =$$

$$= \frac{\operatorname{tr} 1}{N_-(\varepsilon)} = \frac{\dim \mathcal{H}}{N_-(\varepsilon)} \qquad \text{q.e.d.}$$

If $\dim \mathcal{H}$ is given along with $(\Omega,d)$ then (1) gives an upper bound on $\Delta(0)$ for possible embeddings while (2) shows that this bound cannot be improved indefinitely. ($\varepsilon$ occurs only as a free parameter in (2). For example, if $\varepsilon$ is chosen such that $N_-(\varepsilon) \geq 2 \dim H$ then $\Delta(0) \geq \varepsilon/2$). This conclusion may be interpreted as a "thermodynamic uncertainty relation" as it sets limits to the accuracy of macroscopic state descriptions for systems with an underlying microscopic structure.

The following Theorem extends the result (1) to time evolutions:

THEOREM 2

Let $(\Omega,d)$ be a precompact metric space, $V : \mathbb{R}_+ \times L_M \to L_M$ a time evolution which is norm-continuous on $\mathbb{R}^+ \times \Lambda_d$. Let $\varepsilon > 0$ and $0 < \tau < \infty$.

Then there is a finite dimensional Hilbert space $H$ with unitary time evolution $U_t$ and an embedding $(S,T)$ such that

$\Delta(t) \leq \varepsilon$ for all $t \in [0,\tau]$.

Sketch of proof: 1) If $f_1$ and $f_2$ are approximately identical functions on $\Omega$ it does not follow in general that $V_t f_1$ and $V_t f_2$ are also similar. To compensate for this one introduces the metric

$$\tilde{d}(\omega_1,\omega_2) := \sup_{f \in \Lambda_d} \sup_{t \in [0,\tau]} |V_t f(\omega_1) - V_t f(\omega_2)|.$$

Then $(\Omega,\tilde{d})$ is again a precompact metric space by the continuity condition on $V$ and the Arzela-Ascoli theorem. As in theorem 1 one finds finitely many $\eta_n \in K_M$, $h \in L_M$ such that

$$\forall_{\rho \in K_M} \; \forall_{f \in \Lambda_d} \; \forall_{t \in [0,\tau]} \; \left|\sum_n \langle\rho,h_n\rangle\langle\eta_n,V_t f\rangle - \langle\rho,V_t f\rangle\right| < \varepsilon/2.$$

2) By taking direct sums over n the problem is reduces to approximating the functions $\langle\eta_n,V_t f\rangle$ by $\langle W_n,U_t T_n(f)\rangle_{H_n}$ uniformly in $f \in \Lambda_d$ and $t \in [0,\tau]$. We may assume that $V_t$ is periodic (with a period larger than $\tau$). By rescaling the time parameter the period is set to $2\pi$ and it will be convenient to construct $U_t$ with the same period. Since $T_n$ must depend linearly on f it is plausible to define:

$T_n(f) := \int_0^{2\pi} dt \langle\eta_n,V_t f\rangle \, U_{-t}(\theta)$ for some fixed positive operator

$\theta \in B(H_n)$ satisfying $\int_0^{2\pi} dt \, U_{-t}\theta = 1$. With this definition:

$$|\langle\eta_n,V_t f\rangle - \langle W_n,U_t T_n f\rangle_{H_n}| = |\langle\eta_n,V_t f\rangle - \int_0^{2\pi} dt' \langle\eta_n,V_{t'} f\rangle\langle W_n,U_{t-t'}\theta\rangle_{H_n}|$$

$$\leq \int_0^{2\pi} dt' |\langle\eta_n,V_t f\rangle - \langle\eta_n,V_{t'} f\rangle| \cdot \langle W_n,U_{t-t'}\theta\rangle_{H_n}.$$

By the equicontinuity of the family of functions $\{\langle\eta_n, V_{\cdot} f\rangle\}_{f\in\Lambda_d}$ it suffices to prove that the kernel $\langle W_n, U_t\theta\rangle_{H_n}$ may be chosen to be strongly peaked at t = O. This is done by the explicit construction of $H_n$ (dim $H_n$ =: D), $W_n$, $U_t$, $\theta$ with

$$\langle W_n, U_t\theta\rangle_{H_n} = \frac{1}{2\pi D}\left|\frac{\sin\frac{Dt}{2}}{\sin t/2}\right|^2 \quad \text{q.e.d.}$$

The dimension of the Hilbert space constructed in this proof may be estimated similarly as in theorem 1. Roughly speaking dim $H = N_1 \cdot N_2$, where $N_1$ arising from part 1 of the proof, is a covering number $N_+$ for the metric $\tilde{d} \cdot \tilde{d}$ is again a metric of physical imprecision taking into account the additional possibility of distinguishing points of $\Omega$ by measurements performed a long time after preparation. The ratio $d/\tilde{d}$ characterizes the "stability" of the macroscopic motion. This means that a wildly unstable macroscopic theory (like hydrodynamics in the turbulent regime) might not be consistent with a microscopic description (in a Hilbert space of given dimension) and should not be taken too seriously as a physical theory.

The factor $N_2$ arises from the approximation of the functions $\langle\eta, V_t f\rangle$. It depends on $\tau$, $\varepsilon$ and the rate at which these functions vary. (Suppose $\left|\frac{d}{dt}\langle\rho, V_t f\rangle_M\right| \leq \alpha$ uniformly for $\rho \in K_M$, $f \in \Lambda_d$, $t \in [0,\tau]$. Then $N_2$ is of the order $(\alpha\tau\varepsilon^{-1})^2$.) For example if typical macroscopic state changes happen on a scale of microseconds and approximation is assumed to be better than $10^{-6}$ for some million years, $N_2$ will be of the order $10^{50}$.

Large as this may seem, $N_2$ will still be negligible compared to $N_1$ in typical situations: Suppose that the macroscopic motion is stable $(d=\tilde{d})$ and that the given accuracy allows us only to distinguish two different states in a volume of 1 $mm^3$. Then for a total volume of one liter (= $10^6$ $mm^3$):
$\log N_1 \approx 10^6 \log 2 \gg 50 \approx \log N_2$. In other words $\log N_1$ is an extensive quantity (increasing linearly with volume) whereas $\log N_2$ does not depend on the size of the system and will consequently be irrelevant for large enough (i.e. macroscopic) systems.

In order to apply this theorem to a concrete physical system one needs an estimate of dim $H$ which has to be derived from the microscopic theory. This value together with the estimates discussed above will decide how large the thermodynamics uncertainties must be with respect to which macroscopic and microscopic theories are compatible.

Recall that $H$ was supposed to be that portion of the (infinite dimensional) Hilbert space of the system on which the Hamiltonian is bounded by some fixed energy E. By the rules of equilibrium statistical mechanics log dim $H$ is then equal to the entropy S(E) of the equilibrium system with energy E. But this quantity is well known: Entropy is also an extensive quantity and (e.g. for an ideal gas at room temperature) is of the order of the particle number, say $10^{23}$. Comparing this with estimates of $\log N_1$ of the type used above, we conclude that there is indeed enough room in Hilbert space to admit very small thermodynamic uncertainties.

REFERENCES

[1] Ludwig, G: The connection between the objective description of macro-systems and quantum mechanics of many particles. In: A. van der Merve (ed.): Essays in Honor of Wolfgang Yourgrau, Plenum Press, N.Y. 1981 (to be published)
[2] Ludwig, G: A Theoretical Description of Single Microsystems. In: W.C. Price, S.S. Chissick (eds.): The Uncertainty Principle and Foundations of Quantum Mechanics. Wiley & Sons, London 1977
[3] Werner, R: Doctoral dissertation, in preparation
[4] Mayr, D: Article in this volume
[5] Ludwig, G: Article in this volume

# STABLE AXIOMS IN PHYSICAL THEORIES

H.-J. Schmidt

Fachbereich Physik, Universität Osnabrück
Postfach 4469, 4500 Osnabrück
Federal Republic of Germany

## I. INTRODUCTION

Stability of a form may defined intuitively by demanding that small perturbations of the form cause no changes or only "well-behaved" changes of the form. In the words of R. Thom [1]: "Une G-forme A sera dite *structurellement stable*, si toute form B assez voisine de A... est G-équivalente à A...".
(R. Thom considers a pseudo-group G operating on forms.) In physics one often deals with stability of states, e.g. of orbits under small perturbations or of thermodynamical states under external fluctuations [2]. However, one may also examine stability of a physical theory (*PT*) as a whole. The "perturbations" here would correspond to confrontations with the real world or with rival theories, which threaten to "destroy the form", i.e. to falsify the theory. Following G. Ludwig [3] and U. Moulines [4], the *PT* can be stabilized against both kinds of "perturbation" by passing from exact applications to approximate applications and from exact intertheoretical relations to approximate ones. (The technical details involving "imprecision-sets", i.e. entourages of "physical uniformities" are explained in the respective contributions of this volume.)

In this essay we shall be concerned with the question: What kind of stability resp. G-equivalence (in R. Thom's terms) of a *PT* is actually implied by that technique of approximation? In other words, which reformulations of the mathematical theory would not affect the approximate applications of a *PT*? If in this question "approximate" is replaced by "exact", there already exists a solution due to G. Ludwig [3], which results in the concept of an "axiomatic basis" of a *PT* using the metamathematical notion of a "species of structures $\Sigma$". Different mathematical formulations then correspond to isomorphic structures of the same species $\Sigma$. Within the mathematical theory employing $\Sigma(MT_{\Sigma})$ a structure describing physical situations can only be determined "intrinsically" up to $\Sigma$-isomorphisms. If we replace the notion "set" in this framework by "uniform space" and "isomorphism" by "blurred isomorphism" (e.g. the real line and the rational line are blurred isomorphic), we obtain the concept of a "stable species of structures" which plays the analogous role as the original "species of structures" for exact applications of a $PT$. We shall then define "stable terms" analogous to "intrinsic terms" and "strongly stable" species of structures which allow the transport of structues onto blurred-isomorphic sets. Finally we shall discuss possible connections of our concepts to "approximate reduction" of one $PT$ by another.

Before going into details, I would like to sketch the relevance of the first mentioned kind of stability of orbits for a deep problem in the foundations of physics which may be summarized in the question: why are orbits differentiable?

Obviously, differentiability cannot be simply a finding of fact. Perhaps the theory-dependance of the statement of differentiability is best illuminated by the following remark: if a billiard ball thrown into the air is considered as a big Brownian particle, then with probability 1 its orbit is not differentiable (cf. [5] Theorem 2.2). The orbit of a body is not "given" quantitatively. At most a finite number of extended processes in the neighbourhood of the orbit is given. These processes clearly do not determine a unique orbit, they could only be expected to determine a certain neighbourhood in the space of all orbits. Unfortunately the topology induced in this way is not Hausdorff, even if restricted to $C^{\infty}$-orbits. And, what is still more disturbing, differentiation is not a continuous operation under these circumstances. However, these handicaps may disappear, if one restricts the orbits to a very small class, for instance to conic sections or, more generally, to the class of solutions of an equation of motion. Thus it looks as if "orbit" will be a theoretical concept which can only be defined by means of the laws of, say, classical mechanics. Thus its differentiability depends on the kind of law considered. It turns out to be different depending on whether we consider a Markov process or a (deterministic) differential equation. With respect to the latter, the possibility of "orbit-determination" seems to be intimately connected with stability of the differential equation in the usual sense.

## II. STABLE SPECIES OF STRUCTURES

The concept of a species of structures can be used as a framework for the formal reconstruction of physical theories. Since there are detailed accounts on this point [3,6,7,8,9] we may confine ourselves to some sketchy remarks. We will follow N. Bourbaki [9] modulo rigor. Our notations are rather influenced by the reformulations given by E. Scheibe [6,7,8] , including his shorthand notation. Moreover, we will not strictly distinguish mathematical and metamathematical levels of language, and, for example, say simply "A holds" instead "A is a theorem in $\mathcal{MT}_\Sigma$". Hopefully these concessions to readability will not cause any misunderstandings.

Let us begin with an example. Consider a certain version of the extensive system: it consists of a primitive set X, the elements of which represent measuring rods, two predicates Sxy for "x is shorter than y" and Cxyz for "putting together x and y defines a new rod z" and certain axioms. Let the predicates be represented by sets s,c such that

(2.1) $r = \langle s,c\rangle \in \mathcal{P}(X\times X) \times \mathcal{P}(X\times X\times X)$

and the conjunction of the axioms be denoted by $\alpha(X,r)$.

More generally, let $X_1 \dots X_n$ be the primitive sets ('principal base sets'), and L an echelon operator ('echelon construction scheme'); this means that $L(X_1 \dots L_n)$ is a set obtained from $X_1 \dots X_n$ by forming power sets ($\mathcal{P}$) and Cartesian products. Further, let $r = \langle r_1 \dots r_m\rangle$ be a term, subject to the 'typification axiom'

(2.2) $r \in L(X_1 \dots X_n)$

and the 'proper axiom'

(2.3) $\alpha(X_1 \dots X_n, r_1 \dots r_m)$.

The proper axiom is assumed to satisfy a condition of 'transportability', which will be explained below. These data will define a <u>species of structures</u> $\Sigma$. Using the shorthand notation $X = \langle X_1 \dots X_n\rangle$ we shall say that $\langle X,r\rangle$ is a <u>structure</u> of species $\Sigma = \langle L,\alpha\rangle$. L is the <u>type</u> of $\Sigma$. Let us assume further that the mathematical entities $X_1 \dots X_n$ process an interpretation as different sorts of physical objects and the $r_1 \dots r_m$ as physical predicates which may be assigned to (finite sets of) objects. If this interpretation can be achieved without reference to the physical theory $\mathcal{PT}$ under consideration, the formal theory obtained in this way is an 'axiomatic basis' of $\mathcal{PT}^{(1)}$. For the sake of simplicity we do not consider 'auxiliary base sets' as **C** or **R**, thus confining ourselves to qualitative axiom systems, and, secondly, we do not allow for representing a sort of physical objects as an echelon term over primitive sets. A geometrical axiom system, for example, could not treat 'lines' as sets of 'points', but would have to introduce 'lines' and 'points' as independent entities. It seems plausible that every axiomatic basis of a $\mathcal{PT}$ may be reduced to this form, which is simpler to handle in our context.

An <u>application</u> of the $\mathcal{PT}$ consists of a statement that certain predicates apply to certain concrete objects. This can be expressed in the form:

(2.4) $\langle x_{\nu_1} \ldots x_{\nu_k} \rangle \in r_\nu$

or as a compound sentence (without quantifiers) involving such propositions, where the constants $x_{\nu_i}$ are elements or finite subsets of the primitive sets $X_{\nu_i}$. Straining our shorthand notation a bit, we may also write this as:

(2.5) $x \in \bar{r}$ and $\bar{r} \in R(X)$,

where R is an appropriate echelon operator. Since in most cases (2.5) would contradict the axiom $\alpha(X,r)$, it must be weakened, or 'blurred', to the following:

(2.6) $\exists y : xNy$ and $y \in \bar{r}$,

where xNy is some relation of 'nearness' or <u>imprecision-set</u>. Following G. Ludwig [3] , we assume that the class of admissible N can be chosen from so-called <u>uniformities</u> on the primitive sets,

(2.7) $\mathcal{U}_i \in P^2(x_i \times x_i)$, shortly: $\mathcal{U} \in P(X)$.

The uniformities $\mathcal{U}_i$ on the sets $X_i$ may be part of the structural term r. In mature theories, where the inaccuracy is controlled by measurements, so to speak, $\mathcal{U}_i$ may be a term defined over r. Correspondingly, the axioms of uniformity can be a part of $\alpha(X,r)$ or a logical consequence of it. At any case:

<u>D1:</u> $\langle X,r,\mathcal{U}\rangle$ is a called a structure of <u>(MH-) uniformed</u> species $\Sigma = \langle L,\alpha\rangle$, iff
- (i) $\langle X,r\rangle$ is a structure of species $\Sigma$,
- (ii) $\mathcal{U}$ is a term in $MT_\Sigma$, and
- (iii) $(r \in L(X)$ and $\alpha\langle X,r\rangle) \Rightarrow (\mathcal{U} \in P(X)$ and $\beta(X,\mathcal{U}))$,

where $\beta(X,\mathcal{U})$ comprises the axioms (2) of a (<u>m</u>etrizable <u>H</u>ausdorff) uniformity for all $\mathcal{U}_i$ on the sets $X_i$.

For any uniform space $\langle X_i,\mathcal{U}_i\rangle$ one can define its Hausdorff completion $\langle \hat{X}_i,\hat{\mathcal{U}}_i\rangle$ (3). Thus we may ask, whether a term $\hat{r}$ with r dense in $\hat{r}$ exists such that $\langle \hat{X},\hat{r}\rangle$ is again a structure of species $\Sigma$.

This need not be the case, as simple counter-examples show [4].

First, the typification axiom $\hat{r} \in L(X)$ must hold. Let us assume $L = \mathcal{P}\mathcal{S}$, that is: $r \subset \mathcal{S}(X)$, where $\mathcal{S}$ is another echelon operator. Note that $\mathcal{S}(X)$ can also be considered as a uniform space because the product of uniform spaces and the set of subsets of a uniform space can be endowed with a canonical uniformity [5]. Therefore a natural choice of $\hat{r}$ is the closure of r as a subset of $\mathcal{S}(X)$. Now the typification axiom $\hat{r} \subset \mathcal{S}(\hat{X})$ holds if we can prove $\widehat{\mathcal{S}(X)} \subset \mathcal{S}(\hat{X})$, up to some isomorphism. Clearly it suffices to prove this for $\mathcal{S}$ being a product or a power set operator. The first case: $X \hat{\times} Y \cong \hat{X} \times \hat{Y}$, is a standard theorem [6], but $\widehat{\mathcal{P}(X)} \subset \mathcal{P}(\hat{X})$ turns out to be more delicate. Let us recall the definition of the uniformity on $\mathcal{P}(X)$. If $\langle X, \mathcal{U} \rangle$ is a uniform space, $A, B \in \mathcal{P}(X)$ and $N \in \mathcal{U}$, then $A \tilde{N} B \underset{\text{def}}{\Leftrightarrow} A \subset NB$ and $B \subset NA$. The set of relations $\tilde{N}$ obtained in this way forms a fundamental system of entourages of a uniformity $\tilde{\mathcal{U}}$ on $\mathcal{P}(X)$. $\langle \mathcal{P}(X), \tilde{\mathcal{U}} \rangle$ is in general <u>not</u> Hausdorff, but its subspace $\mathcal{F}(X)$, the set of all <u>closed</u> subsets of X, is. We have the following result [7]:

<u>T1:</u> If X is a metrizable Hausdorff uniform space, then $\widehat{\mathcal{P}(X)} \cong \mathcal{F}(\hat{X})$.

This isomorphism is "canonical" in some sense, which will become clear in the run of the proof.

<u>Proof</u>: It is enough to show that $\mathcal{F}(\hat{X})$ is complete, because $\mathcal{F}(X)$ is the Hausdorff uniform space associated with $\mathcal{P}(X)$ (cf. [12, II § 3.8]) and is dense in $\mathcal{F}(\hat{X})$. Let $(A_i)_{i \in \mathbb{N}}$ be a Cauchy sequence in $\mathcal{F}(\hat{X})$, $\mathcal{U}$ the neighbourhood filter of $\hat{X}$ and define

$\tilde{A} = \{x \in \hat{X} \mid \forall N \in \mathcal{U}\ \exists\, n \in \mathbb{N}\ \forall\, i \geq n : Nx \cap A_i \neq \emptyset\}$.

Clearly, $\tilde{A}$ is closed. Let us show $\lim A_i = \tilde{A}$.
$\forall N \in \mathcal{U}\ \exists\, n \in \mathbb{N}\ \forall i \geq n : \tilde{A} \subset NA_i$ holds by definition of $\tilde{A}$. In order to prove the other direction, $A_i \subset N\tilde{A}$, let $N \in \mathcal{U}$ be given. Choose a closed symmetric neighbourhood $N_1$ such that $N_1^3 \subset N$, and choose $n \in \mathbb{N}$ such that $j \geq n$ implies $A_j \subset N_1 A_n$ and $A_n \subset N_1 A_j$. The latter is possible because $(A_i)_{i \in \mathbb{N}}$ is a Cauchy sequence. Hence, if $i \geq n$ and $x \in A_i$, there exists an $x_1 \in A_n$ such that $x N_1 x_1$.
Now choose a base $\{N_1, N_2, \ldots\}$ of closed symmetric neighbourhoods such that $N_{i+1}^2 \subset N_i$ and choose $n_i \in \mathbb{N}$ such that all $A_j$ for $j \geq n_i$ are $N_i$-close (again using that $(A_i)_{i \in \mathbb{N}}$ is Cauchy).
Thus we can find $x_{i+1} \in A_{i+1}$ $(i = 1, 2, \ldots)$ satisfying $x_{i+1} N_i x_i$. Note that $(x_i)_{i \in \mathbb{N}}$ forms a Cauchy sequence in $\hat{X}$, say $x_i \to \tilde{x}$, and that $xN\tilde{x}$ holds. Further, $\tilde{x} N_i^2 x_i$ and $x_i N_i y$ for some $y \in A_j$ and all $j \geq n_i$. Hence $j \geq n_i$ implies $N_i^3 \tilde{x} \cap A_j \neq \emptyset$. We may conclude that $\tilde{x} \in \tilde{A}$, which completes the proof. □

The restriction to Hausdorff metrizable uniformities is not very strong since these properties are usually ascribed to uniformities describing physical imprecision[8].

According to the previous remarks, T1 entails the typification axiom $\hat{r} \in PS(\hat{X})$. Let $\hat{\alpha}(\tilde{X},\tilde{r})$ denote the axiom: there exists a structure $\langle X,r \rangle$ of species $\langle PS,\alpha \rangle$ such that $\langle \tilde{X},\tilde{r} \rangle$ is isomorphic to the completion of $\langle X,r \rangle$ w.r.t. the term $\mathcal{U}$. Then the following theorem is immediate:

T2: Let $\langle X,r,\mathcal{U} \rangle$ be a structure of MH-uniformed species $\langle PS,\alpha \rangle = \Sigma$ and $\langle \hat{X},\hat{r},\hat{\mathcal{U}} \rangle$ its completion. Then $\langle \hat{X},\hat{r},\hat{\mathcal{U}} \rangle$ is a structure of MH-uniformed species $(PS,\hat{\alpha}) = \hat{\Sigma}$.

Next we want to argue that the theories $PT_\Sigma$ and $PT_{\hat{\Sigma}}$ are physically equivalent by reason of the approximate character of its applications. Let N be an entourage of X (shorthand notation), $N^R$ the associated entourage of R(X) (cf. (2.5)) and

(2.8) $\exists y : xN^R y \text{ and } y \in \overline{r}$

a (blurred) application of $PT_\Sigma$. The corresponding (blurred) application of $PT_{\hat{\Sigma}}$ will be

(2.9) $\exists y : x\hat{N}^R y \text{ and } y \in \overline{\hat{r}}$ .

T3: If (2.8) is consistent with $\alpha(X,r)$, then (2.9) is consistent with $\hat{\alpha}(\hat{X},\hat{r})$.
Conversely, if (2.9) is consistent with $\hat{\alpha}(\hat{X},\hat{r})$ and M is any entourage on X, then

(2.10) $\exists y : x(MN)^R y \text{ and } y \in \overline{r}$

is consistent with $\alpha(X,r)$. (Here MN denotes the "relational product".)

The proof is straightforward and left to the reader.

We thus may summarize, that any mathematical difference between $MT_\Sigma$ and $MT_{\hat{\Sigma}}$ is physically irrelevant, because every (blurred) application of the corresponding physical theory is insensitive to these differences. In other words: the axiom $\hat{\alpha}(\hat{X},\hat{r})$ of $\hat{\Sigma}$ contains some degree of idealization which in principle cannot be justified empirically. Of course, it might be justified by other arguments, say the demand of mathematical elegance or convenience. But for the purpose of methodological analysis, it is neccessary to be able to suspend idealizations, as far as possible. In this context this means that the axiom of the species of structures should be insensitive w.r.t. completion of the primitive sets and structural terms:

D2: Let $\langle X,r,\mathcal{U} \rangle$ and $\langle Y,s,\mathcal{V} \rangle$ be two structures of MH-uniformed species $\langle PS,\beta \rangle$. $\langle X,r,\mathcal{U} \rangle$ and $\langle Y,s,\mathcal{V} \rangle$ are called blurred-isomorphic, $\langle X,r,\mathcal{U} \rangle \overset{\mathcal{E}}{\approx} \langle Y,s,\mathcal{V} \rangle$, iff there exists an isomorphism

$<\hat{X},\hat{r},\hat{\mathcal{U}}> \overset{f}{\to} <\hat{Y},\hat{s},\hat{\mathcal{U}}>$.

D3: An MH-uniformed species of structures $\Sigma = <\mathcal{PS},\alpha>$ is called stable, iff for any two structures $<X,r,\mathcal{U}>$ and $<Y,s,\mathcal{V}>$ of some MH-uniformed species $<\mathcal{PS},\beta>$, which are blurred-isomorphic, $\alpha(X,r)$ implies $\alpha(Y,s)$. Sometimes the axiom $\alpha$ will be called stable too.

D4: An MH-uniformed species of structures $\Sigma = <\mathcal{PS},\alpha>$ is said to be stabilized by another MH-uniformed species $\Sigma_{st} = <\mathcal{PS},\gamma>$ of the same type, iff for all structures $<X,r,\mathcal{U}>$ of the species $\Sigma$ there exists a structure $<X',r',\mathcal{U}'>$ of the species $\Sigma_{st}$ such that $<X,r,\mathcal{U}> \approx <X',r',\mathcal{U}'>$.

The condition, that $\alpha(x,r)$ is conserved under isomorphisms, is just the transportability requirement, which is constitutive for species of structures. Stable species of structures are thus defined to satisfy the stronger condition, namely that its axioms are conserved under "blurred-isomorphisms".

It is rather plausible that all species of structures which are usually considered are not stable, hence this definition doesn't look very reasonable at first sight. On the other hand any species of structures $\Sigma$ may be stabilized in an almost trivial manner, if we formulate as an axiom of $\Sigma_{st}$, that all its structures are blurred-isomorphic to some structure of $\Sigma$.

But we are rather looking for axioms of $\Sigma_{st}$, which have an intuitive and operational meaning(9). Due to this constraint, which is not yet formulated in exact terms, stabilization becomes a non-trivial task. We shall give an example.

## III. AN EXAMPLE: BLURRED ORDER

Consider the species $\Sigma$ of ordered, MH-uniform spaces. If $<X,G,\mathcal{U}>$ is a structure of this species, its completion $<\hat{X},\hat{G},\hat{\mathcal{U}}>$ need not satisfy the axioms of order. (Take for instance G equal to $\leq$ for rationals and equal to $\geq$ for irrationals). Hence the above mentioned trivial $\Sigma_{st}$ would be far too big. But let us examine the axioms of order from a physical point of view, adopting a certain interpretation of X as a set of (classes of reproductions of) "objects" and of xGy as "x is greater/equal than y (w.r.t. some property)". Then

(3.1) "$\forall x \in X : xGx$"

means that each object is greater/equal than any of its reproductions. Taking into account the moment of inaccuracy, say some disturbing influence on the processes of reproduction and comparison, which may be represented by an imprecision-set $N \in \mathcal{U}$,

(3.1) could be regarded as an idealization of:

$$(3.2)\ \forall x \in x\ \exists\ x_1, x_2 \in X : x_1 N x,\ x N x_2 \text{ and } x_1 G x_2.$$

This version depends on the constant entourage N, but if one does not really know the size of N, one is led to the supposition, that N can be made arbitrarily small:

$$(3.3)\ \forall N \in \mathcal{U},\ x \in X\ \exists\ x_1, x_2 \in X : x_1 N x,\ x_2 N x \text{ and } x_1 G x_2.$$

In some respect, this is still an idealization, which however accounts for the moment of inaccuracy. In a slightly more concise way, (3.1) is rewritten as

$$(3.4)\ \Delta \subset G,$$

and (3.3) as

$$(3.5)\ \forall N \in \mathcal{U} : \Delta \subset NGN,$$

or, if $\bar{G}$ denotes the closure of G w.r.t. the topology on $X \times X$ induced by $\mathcal{U}$, as

$$(3.6)\ \Delta \subset \bar{G}.$$

Similiarly, the axiom of asymmetry

$$(3.7)\ G \cap G^{-1} \subset \Delta$$

is modified to

$$(3.8)\ \forall N \in \mathcal{U}\ \exists\ M \in \mathcal{U} : MGM \cap MG^{-1}M \subset N,$$

which, by the way, implies:

$$(3.9)\ \bar{G} \cap \overline{G^{-1}} \subset \Delta.$$

Transitivity,

$$(3.10)\ GG \subset G,$$

becomes:

$$(3.11)\ \forall N \in \mathcal{U}\ \exists\ M \in \mathcal{U} : MGMGM \subset NGN.$$

Again, this implies:

$$(3.12)\ \bar{G}\,\bar{G} \subset G.$$

We remark that the modified axioms (3.5), (3.8), (3.11) are not logically weaker nor stronger than the original axioms of order (3.4), (3.7), (3.10). In some respect, they are weaker, but on the other hand they express a kind of compatibility between $\mathcal{U}$ and G, which is not automatically satisfied. Let us call <X,G,$\mathcal{U}$> a blurred-ordered, MH-uniform space, if the axioms of order are replaced by (3.5), (3.8), (3.11), and denote by $\tilde{\Sigma}$ the corresponding species of structures.

T4: An MH-uniform space <X,G,$\mathcal{U}$> is blurred-ordered iff its completion <$\hat{X},\hat{G},\hat{\mathcal{U}}$> is an ordered space.

Proof: If we identify X with a dense subset of $\hat{X}$, we have $\overline{G} = \hat{G}$ and (3.6), (3.9), (3.12) show, that <$\hat{X},\hat{G},\hat{\mathcal{U}}$> is an ordered space, if <X,G,$\mathcal{U}$> is blurred-ordered. Conversely, (3.6) implies (3.5). Clearly, $(MGM \cap MG^{-1}M)_{M\in\hat{\mathcal{U}}}$ resp. $(MGMGM)_{M\in\hat{\mathcal{U}}}$ are Cauchy-nets in $\mathcal{P}(\hat{X}\times\hat{X})$, thus converging towards Y resp. Z (cf. T1).

We have to prove $Y \subset \Delta$ and $Z \subset \hat{G}$. Suppose first $(x,y) \in Y$. It follows that $(X,Y) \in \overline{MGM}$ for all $M \in \hat{\mathcal{U}}$, hence $(x,y) \in \overline{G}$ and, analogously, $(x,y) \in \overline{G^{-1}} = \overline{G}^{-1}$. Because $\overline{G}$ is an ordering, this implies $(x,y) \in \Delta$. Now consider the case $(x,y) \in Z$. If $\overline{G}x \cap \overline{G}^{-1}y \neq \emptyset$, we are done, thus let us assume that the closed sets $\overline{G}x$ and $\overline{G}^{-1}y$ are disjoint. Since $\hat{X}$ is normal, MGMx and $MG^{-1}My$ must be disjoint for some $M \in \hat{\mathcal{U}}$. This contradicts $(x,y) \in MGMGM$. □

As a corollary we have:

T5: $\tilde{\Sigma}$ is stable.

## IV. STABLE TERMS

When dealing with a structure <X,r> of species Σ one often considers terms Y(X,r), s(X,r) such that <Y,s> is another structure of species Π. For example, the set of automorphisms AUT(X,r) of any structure <X,r>, together with the composition "o" of maps, forms a group.

This leads to the concept of a "procedure of deduction" $\delta: \Sigma \to \Pi$[10]. Intuitively, we may think of δ as a mapping between classes of structures. Thus it is natural to ask, whether $\delta^{-1}[\Pi']$ is a species of structures, if Π' is a species finer than Π. To put it in more familiar terms: let β'(Y,s) be the transportable axiom of Π', is then the axiom β'(Y(X,r),s(X,r)) transportable w.r.t. the typification of r? The simplest way to insure this is to postulate a certain covariance property for the terms Y(X,r),s(X,r): namely that any isomorphism $f: (X_1,r_1) \to (X_2,r_2)$ maps $Y(X_1,r_1)$ onto $Y(X_2,r_2)$, when it is lifted to the appropriate echelon terms, equally for s(X,r). This property defines intrinsic terms, which also should be used if defining physical concepts within a theory $PT_\Sigma$ (11).

The foregoing considerations can easily be carried over to stable species of structures and blurred-isomorphisms. One then obtains the concept of stable terms, which are insensitive to the choice of dense subsets.

D5: Let $\langle X,r,\mathcal{U}\rangle$ be a structure of stable species $\Sigma = \langle PS,\alpha\rangle$, $\tau(X)$ an echelon term, $V(X,r) \subset T(X)$ a term and $\mathcal{V}$ an MH-uniformity on $T(X)$. Let $\hat{V}(X,r)$ denote the $\mathcal{V}$-completion of $V(X,r)$. $V(X,r)$ is called stable iff $\langle X,r\rangle \underset{\approx}{\overset{f}{}} \langle Y,s\rangle$ implies $f^{PT}\hat{V}(X,r) = \hat{V}(Y,s)$. Unless otherwise stated, $\mathcal{V}$ will be the uniformity induced by $\mathcal{U}$.

Consider for example a structure $\langle A,+,e,\mathcal{U}\rangle$ of MH-uniformed species $\Sigma$, such that $\alpha(A,+,e,\mathcal{U})$ is equivalent to the existence of an isomorphism $f: \langle\hat{A},\hat{+},e,\hat{\mathcal{U}}\rangle \rightarrow \langle\mathbb{R}_+,+,1,\mathcal{V}\rangle$, $\mathcal{V}$ being the usual uniformity on $\mathbb{R}_+$. Clearly, $\Sigma$ is then stable. Let a function $f_A: A \rightarrow \hat{A}$ be an intrinsic term, and denote by $\hat{f}_A \subset \hat{A} \times \hat{A}$ the completion of its graph. Then $f_A$ is stable iff for any dense subset $B \subset \hat{A}$ the corresponding $\hat{f}_B$ is equal to $\hat{f}_A$. Hence a continuous function is stable, but a function, which is continuous up to a nowhere dense subset, provided that it is appropriately defined on that subset, is also stable.

The set of connected components (12) of a uniform space is another example of an intrinsic term which is not stable in general.

We shall define the notion of a stable procedure of deduction and collect some obvious results.

T6: Let $\langle X,r,\mathcal{U}\rangle$ be a structure of stable species $\Sigma$. Then the terms $\hat{X}$, $\hat{r}$, $\hat{\mathcal{U}}$ will be stable. If $V(X,r,\mathcal{U})$ is an intrinsic term, then $V(\hat{X},\hat{r},\hat{\mathcal{U}})$ will be stable in $\Sigma$.

For the proof note, that the usual uniformity of $\hat{X}$ coincides with its canonical uniformity as a subset of $PP(X)$ (namely the subset of minimal Cauchy filters).

D6: Let $\Sigma,\Pi$ be MH-uniformed species of structures, $\Sigma$ be stable. A procedure of deduction $\delta: \Sigma \rightarrow \Pi$ will be called stable if the corresponding terms $Y(X,r)$, $s(X,r)$ are stable in $MT_\Sigma$.

T7: The product $\varepsilon\delta$ of two stable procedures of deduction $\Sigma \overset{\delta}{\rightarrow} \Pi \overset{\varepsilon}{\rightarrow} \Theta$ (which is defined in the obvious way) is a stable procedure of deduction.

T8: Let $\delta: \Sigma \rightarrow \Pi$ be a stable procedure of deduction, $\Pi'$ a stable species of structures which is finer than $\pi$. Then $\delta^{-1}[\Pi']$

is a stable species of structures, finer than $\Sigma$. More explicitely: If $\beta'(Y,s)$ is the axiom of $\Pi'$ and $\alpha(X,r)$ the axiom of $\Sigma$, then the axiom "$\alpha(X,r)$ and $\beta'(Y(X,r), s(X,r))$" is stable.

T9: Let $\Sigma$ be a stable species of structures with axiom $\alpha(X,r)$ and let $\beta(X,r)$ be another transportable axiom. Then the axiom "$\alpha(X,r)$ and $\beta(\hat{X},\hat{r})$" will be stable.

## V. STRONG STABILITY

Transportability of an axiom entails the possibility to "transport structures": if $r \in L(X)$ and $\alpha(X,r)$ and $f: X \to Y$ is a bijection, then there exists an $s \in L(Y)$ such that $\alpha(Y,s)$ holds and $f: \langle X,r\rangle \to \langle Y,s\rangle$ will be an isomorphism, namely $s = f^{L}r$.

The analogous result for blurred-bijections $X \underset{\approx}{f} Y$ and stable axioms is not generally valid.

Take for example the (stable) species of all MH-uniform spaces with addition, which are blurred-isomorphic to $(\mathbb{R}_+,+)$. There exist dense subsets $F \subset \mathbb{R}_+$, on which the addition $+$ cannot be densely defined. In other words, if $\alpha \subset F \times F \times F$, hence $\hat{\alpha} \subset \mathbb{R}_+ \times \mathbb{R}_+ \times \mathbb{R}_+$, then $\hat{\alpha} \neq +$. Such an $F$ can be explicitely defined as follows: Let $n \in \mathbb{N}$ and choose a prime number $p_n$ from the interval $(\sqrt{n}\, 2^n, \sqrt{n+1}\, 2^{n+1})$ and consider the set $F_n$ of all fractions of the form $2^n(k+2^{-k})/p_n$, where $k \in \mathbb{N}$ and $k \leq n$, which are not integers.

Let $F = \bigcup_{n\in\mathbb{N}} F_n$. The fractions in $F_n$ cover an interval of order $2^n n/p_n \simeq \sqrt{n}$ and their spacing is of order $2^n/p_n \simeq 1/\sqrt{n}$. Hence $F$ is dense in $\mathbb{R}_+$. But one may easily establish that $x,y \in F$ implies $x + y \notin F$, hence $\alpha \cap + = \emptyset$ and $\hat{\alpha} \neq +$.

We may thus define a species of structures $\Sigma$ to be "strongly stable" if that kind of "blurred-transportability" is satisfied.

D7: An MH-uniformed species of structures $\Sigma = \langle L,\alpha\rangle$ is called strongly stable, iff for each structure $\langle X,r,\mathcal{U}\rangle$ of $\Sigma$ and each MH-uniform space $\langle Y,\mathcal{V}\rangle$ satisfying $\langle X,\mathcal{U}\rangle \underset{\approx}{f} \langle Y,\mathcal{V}\rangle$ there exists a term $s \in L(Y)$ such that $\langle X,r,\mathcal{U}\rangle \underset{\approx}{f} \langle Y,s,\mathcal{V}\rangle$ and $\alpha(Y,s,\mathcal{V})$ holds.

We have the following sufficient condition for a stable species of structures to be strongly stable.

D8: A subset $C$ of a uniform space will be called full, iff the interior of $\hat{C}$ is dense in $\hat{C}$.

T10: A stable species of structures $\Sigma = \langle PS,\alpha\rangle$ is strongly stable, if for each structure $\langle X,r\rangle$ of $\Sigma$, $r$ is full as a subset of $S(X)$.

For the proof it is easily shown that "Y is dense in $\hat{X}$" implies $s \underset{def}{=} \hat{r} \cap S(Y)$ is dense in $\hat{r}$" □

The afore-mentioned species of blurred-ordered spaces therefore becomes strongly stable if one adds an axiom "$G \subset X \times X$ is full". Of cource, there are blurred-ordered spaces where G is not full.

Axioms, which are not stable, indicate a certain degree of idealization present in the laws of the theory. Axioms, which are stable, but not strongly stable, are rather characteristic of idealization present in the basic concepts. Therefore "strong stabilization" of axioms would be a means to suspend this kind of idealization.

## VI. STABILITY AND APPROXIMATE REDUCTION

Stabilization was proposed as an instrument to distinguish between the idealized and the non-idealized constituents of a theory. The idealized constituents of a theory, even of a successful theory, may possibly not survive the rise of a more comprehensive and more successful theory, because they can be regarded as something "added" to the laws of nature by physicists. Thus there should be some connection between stability and approximate reduction. However the definition of the latter within Ludwig's framework is not in a final form and therefore I must confine myself to some general remarks and conjectures.

At first glance, also stable axioms can be rejected by later theories.

Roughly speaking, this means that the domain of successful applications, of the physical theory $PT_1$ is equally explained by $PT_2$, and even better, i.e. using smaller imprecision - sets Moreover, the axioms of $PT_1$ are not consistent with those of $PT_2$, but can be deduced from $PT_2$ "within some degree of approximation". The later theory $PT_2$ therefore <u>explains</u> a certain portion of the imprecision - sets, which were needed to immunize $PT_1$ against falsification. $PT_2$ thus determines the limits of the domain of successful applications of $PT_1$ more accurately that before. Therefore it is likely that the axioms of $PT_1$ will survive in some approximate form. This approximate form is <u>not</u> the stable form, it cannot be, as long as physics does not rest on prophesy. But I conjecture, that the surviving approximate form results from the stable form simply by suspending the remain of idealization, that is by restricting the class $\mathcal{U}$ to finite imprecision-sets. This conjecture will be substantiated by a simplified example.

Consider in $PT_1$ a structure $\langle X, \mathcal{U}, G \rangle$ subject to the axioms of MH-uniformity and blurred-transitivity (3.11). Think of the elements of X as representing measuring rods and of G as representing the comparison in length. The more comprehensive theory

$PT_2$ tells us that every rod $x \in X$ has a "true length" $l(x)$, which will be shrinking by the amount $L(x,y)$, if x is brought into contact with $y \in X$ for the sake of comparison. Hence xGy will be equivalent to

$$(6.1) \quad l(x) - L(x,y) \geq l(y) - L(y,x).$$

It is clear that (3.11) cannot be true for arbitrarily small $N \in \mathcal{U}$ because of the finite "error of measurement". Let

$$(6.2) \quad \tilde{\Box} \underset{\text{def}}{=} \{(x,y) \in X \times X \mid 0 \leq \frac{l(x) - l(y)}{L(x,y) - L(y,x)} \leq 1\}$$

and $\Box$ be the smallest set which contains $\tilde{\Box}$ and satisfies "$x \Box y$ and $(l(x) \leq l(z) \leq l(y)$ or $l(y) \leq l(z) \leq l(x))$ implies $z \Box y$". Further let $\mathcal{U}_\Box$ be the subclass of neighbourhoods in $\mathcal{U}$ which contain $\Box$. The following axiom is equivalent to blurred-transitivity:

$$(6.3) \quad \forall N \in \mathcal{U}\ \exists\ M \in \mathcal{U} : GMG \subset N^3GN^3.$$

It is still false in $PT_2$, but its approximate version

$$(6.4) \quad \forall N \in \mathcal{U}_\Box\ \exists\ M \in \mathcal{U}_\Box : GMG \subset N^3GN^3$$

can be proved in $PT_2$. (Note, that equivalent forms of an axiom may well possess non-equivalent approximations.) The flexibility needed to account for all possible future corrections just results from the axioms of uniformity. For instance, we have used the "$N^2 \subset M$"-law to obtain the form (6.3) of the stable axiom, which in its modified form (6.4) reflects the propagation of error pertaining to our "theory of measurement" $PT_2$. The stable form of an axiom seems to represent all we can know about its limits of validity, without going beyond the particular theory.

One might object, that our example is untypical in so far, as the concept of length is conserved, whereas in the history of science concepts may lose their legitimation, e.g. "path of a particle" in quantum theory. As pointed out in the introduction, the notion of a path must be viewed as a highly idealized concept, and thus it is indeed a likely victim of scientific progress. Hence, in order to substantiate our conjecture, we would have to offer a criterion for the degree of idealization of concepts. For the time being we can only conjecture that strongly stable theories contain concepts with a "small" degree of idealization and that strongly stable axioms essentially remain valid even if science evolves into more comprehensive theories.

NOTES

(1) All primitive terms of the axiomatic basis correspond roughly to "non-theoretical concepts" in the work of J. Sneed [10],

but not conversely. Note, that the axiomatic basis of a *PT* may be obtained in a somewhat trivial way by means of Ramsey sentences, if a set of non-theoretical concepts in a $MT_\Sigma$ is already singled out [11]. On the other hand, the construction of an axiomatic basis with first order axioms may be an extremely difficult task.

(2) Cf. [12, II § 1.1 and IV § 2.4] or the contributions of G. Ludwig and U. Moulines in this volume.

(3) Cf. [12, II § 3.7].

(4) The norm-completion of the general linear group in n-dimensions is not a group.

(5) Cf. [12, II § 2.6 and § 1 Ex. 5] or the contribution of D. Mayr in this volume.

(6) Cf. [12, II § 3.9].

(7) The following condition can be weakened to the necessary and sufficient condition:"$\hat{X}$ is paracompact and every locally fine modification of $\hat{X}$ is fine", cf. [13, Sect. 41 Ex. 5a]. The case of bounded metric $\hat{X}$ is covered by [12, IX § 2, Ex. 6].

(8) Cf. [3, p. 110].

(9) The analogous situation for the axiomatic basis of a *PT* is mentioned in note (1).

(10) Cf. [9, IV § 1.6].

(11) We provide two examples. In the axiomatic basis of physical geometry according to [14], the quotient of distances is an intrinsic term, the distance itself is not. In an axiomatic basis of (classical or) quantum electrodynamics the spin of the photon would not be gauge-invariant [15, 2.6 and 2.8], hence it would not be an intrinsic term.

(12) Cf. [12, I § 11.5].

REFERENCES

[1] R. Thom: Stabilité Structurelle et Morphogénèse. Benjamin, Reading 1972.
[2] R. Haag, D. Kastler, E.B. Trych-Pohlmeyer: Stability and Equilibrium States. Commun. math. Phys. 38, 173-193 (1974).
[3] G. Ludwig: Die Grundstrukturen einer physikalischen Theorie.

Springer, Berlin 1978.
[4] C.U. Moulines: Approximative Explanation of Empirical Theories: A General Explication. Erkenntnis 10, 201-229 (1976).
[5] T. Hida: Brownian Motion. Springer, New York 1980.
[6] E. Scheibe: Theorien, Strukturarten und mengentheoretische Prädikate. Preprint 1976.
[7] E. Scheibe: On the Structure of Physical Theories. In: The Logic and Epistomology of Scientific Change. Acta Phil. Fenn. XXX (1978), Issues 2-4. 205-224.
[8] E. Scheibe: Invariance and covariance. To appear in: Festschrift for Mario Bunge. Ed.: J. Agassi and R. Cohen.
[9] N. Bourbaki: Theory of Sets. Hermann, Paris 1974 (2nd printing).
[10] J. Sneed: The Logical Structure of Mathematical Physics. Reidel, Dordrecht, 1971.
[11] G. Ludwig: Axiomatische Basis einer physikalischen Theorie und theoretische Begriffe. Preprint 1980.
[12] N. Bourbaki: General Topology. Hermann, Paris 1966.
[13] E. Čech: Topological spaces. Interscience, London, 1966.
[14] H.J. Schmidt: Axiomatic characterization of physical geometry. Lecture Notes in Phys. 111, Springer, New York 1979.
[15] J.M. Jauch, F. Rohrlich: The Theory of Photons and Electrons. Springer, New York 1976.

# APPROXIMATIVE REDUCTION BY COMPLETION OF EMPIRICAL UNIFORMITIES

## Kepler's theory of planetary motion and Newton's theory of gravitation

D. Mayr

Fachbereich Physik der
Philipps-Universität Marburg
Renthof 7, 3550 Marburg
Federal Republic of Germany

## I. INTRODUCTION

In physics almost all important examples of theory reduction are of an approximative nature, for instance the reductions of nonrelativistic to relativistic theories of spacetime and dynamics, or of classical to corresponding quantum theories. To formalize the idea of theory approximation one needs a mathematical structure with a certain type of convergence - the Cauchy convergence, which is defined on a uniform structure (comp. [9], [10], [11]). This convergence type reflects the general fact that the approximated structures or models of the reduced theory are not models of the reducing theory, i.e. the approximation process takes you out of the approximating theory, as it were. For instance in the well-known reduction example of Kepler's theory of planetary motion to Newton's theory of gravitation, Keplerian orbits are not solutions of Newton's differential equations.

But approximative reduction is not a mere mathematical tool for comparing structures of physical theories. A deeper understanding of theory approximation in terms of comparability and progress requires a more empirical characterization of the approximation procedures. One has to use empirical imprecision-sets[1)], characterized by real measurements (comp. [9], § 6), to distinguish the domains in which the reduced and the reducing theory are practically equivalent, in which the reducing theory agrees more exactly with measurements, or in which the reduced theory fails to describe reality. Empirical imprecision-sets are taken into account formally by so-called empirical uniformities which are defined on the fundamental sets of the theory (comp. [3] IV;

[9] § 6). For purposes of approximation the empirical uniformities have to be extended canonically to the (echelon-) sets upon which the structures (relations and functions) of the theory are defined.

The introduction of empirical uniformities also concerns a conceptual distinction of approximation, due to the fact that physicists and mathematicians use the concept of approximation in different ways. In mathematics, for instance, a sequence approximates a (limit-) element of a set, if there are members of the sequence which are as close to the limit as you please, i.e. the usual concept of convergence. However, in physics a further conception is in use; the neglect of secondary effects - which are assumed to be small and will be taken into account later - is the corresponding property in that physical approach. For this kind of approximation it is sufficient that the members of a sequence are only "rather close to the limit", i.e. physical approximation really concerns more a concept of "proximation". This is expressed by the fact that physicists set no bounds to the admissible inaccuracies - much to the confusion of mathematicians (comp. [9] § 8, p. 99). Obviously, the problem is to specify the concept of proximation in terms of mathematical approximation relations. In the following we will treat this delicate question only in rough outlines (chapt. II).

To illustrate the various abstract concepts of theory approximation we will work with the above mentioned reduction example "Kepler-Newton". To begin with, however, the following remarks[2)] will deal with some possible objections. In general, physicists may call Kepler's theory pre-physical, because there is no dynamics in the theoretical framework (although we will present it in a formal guise which suggests some kind of Newtonian dynamics). Thus, physicists may talk only with reservation about a Keplerian limit-case of Newton's theory of gravitation. In particular, Newtonian systems with more than two particles are presumably more or less ergodic, but Keplerian systems are not ergodic in any case. Moreover, the usual approximation of planetary orbits by letting planet masses go to zero depends not only on the quasi-periodic influences of the planet-planet interactions. More important are the secular perturbations which complicate the approximation.

A simple model may explain these considerations. Suppose a planetary model in which two pairs of neighboring planets move around a sun in a circular orbit[3)]. After a sufficiently long time the planet-planet interaction causes the planets to alter their distances and orbits - in contradiction with Kepler's laws of planetary motion. Obviously there are no Newtonian models which approximate that Keplerian roundabout model in the usual distance approximation for all times. Consequently, if one wants to treat the "Kepler-Newton" case as an example of approximative reduction, one has to counter these objections with arguments concerning commensurability and progress. To emphasize this line of reasoning

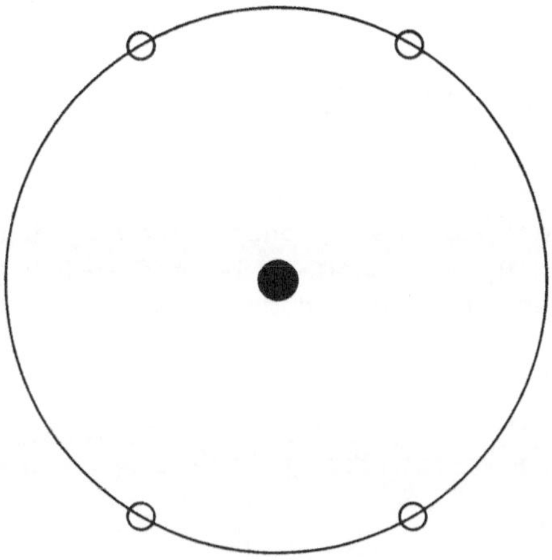

Fig. 1. A special roundabout model

we prove Kepler's theory to be a limit-value of Newton's theory. The approximation holds for empirical uniformities not finer than the uniformity of compact convergence.

## II. FORMAL CONCEPTS OF APPROXIMATIVE THEORY REDUCTION; LUDWIG'S STRUCTURALISTIC APPROACH, PROXIMATION AND APPROXIMATION

The theory concept, used in the following, is modelled on Ludwig's approach to physical theories (comp. [9]). We shall not assume that the reader is completely familiar with the main tool of this structuralistic conception, the so-called species of structures. But within the scope of our essay we cannot present a very detailed reconstruction. Thus, reference to the basic elements of Ludwig's view will have to suffice.

Roughly, a physical theory is a triple consisting of a mathematical theory, a specified region G in reality and certain maps from the elements of G into "base sets" of the mathematical theory - the so-called "Abbildungsprinzipien". The basic term of the mathematical theory is the concept of species of structures (comp. [3] IV, § 1.4), representing in the mathematical framework some physical structures which may be found in the domain G. Species of structures are defined on some base sets $B_1,\ldots,B_n$ whose elements are interpreted, for example, as particles, space and time points, bodies, processes etc. From the base sets, new sets are constructed by the so-called echelon construction schemes ([3] IV, § 1.1), i.e. one works sucessively with the operations P (power-set) and × (product-set).

For instance

$$S_1(B_1 \ldots, B_n) := P(B_1 \times B_2)$$

$$S_2(B_1, \ldots, B_n) := B_1 \times P(B_2)$$

are two different echelon sets constructed by the schemes $S_1$ and $S_2$. The sets $B_1, \ldots, B_n$ are said to be endowed with the structure s iff there is an echelon construction scheme S with

$$s \in S(B_1, \ldots, B_n)$$

(type characterization) and a relation

$$R(B_1 \ldots, B_n; s)$$

which specifies some properties of s. The type characterization $T(B_1, \ldots, B_n; s)$ and the relation $R(B_1, \ldots, B_n; s)$ are called the axioms of the species of structures $\Sigma$. For example the structure d of a metric on the base set B is characterized by

(T) $d \in P(B \times B \times \mathbb{R})$

(R) d is the graph of a function $B \times B \rightarrow \mathbb{R}$ which fulfills the usual properties of a metric.

The set $\mathbb{R}$ of real number is often used as an auxiliary base set. The set $\sigma$ of all elements s which satisfy the axioms T and R is said to be set of structures of the species $\Sigma$ on $B_1, \ldots, B_n$. Finally, the fundamental laws of a physical theory are represented by some relations $\mathcal{A}$ (axioms) on the set of structures. In the following we will characterize the mathematical part of a physical theory by the tuple[4)]

$$(B_1, \ldots, B_n; \sigma_1, \ldots, \sigma_m; \mathcal{A}).$$

To illustrate these rather abstract concepts we return to our reduction example, i.e. to the physical theories of Keplerian motion of planets and Newtonian gravitation. For reasons of simplicity we shall rely on Scheibe's meticulous approach (comp. [13], p.102 ff) and present both theories without regard to historical circumstances.

To begin with, we observe that Kepler's and Newton's theory utilize the same kinematics. Both theories describe paths of particle systems which are assumed to move in a Newtonian spacetime. Obviously, these common features may be restated by a common class of base sets $\mathcal{J}$, $\Theta$ and X, whose elements are interpreted as particles, time and space points. As usual we assume further structures on the sets X and $\Theta$ such that X becomes a 3-dimensional euclidian space; $\Theta$ is endowed with a timelike distance structure (comp. [9], p. 75), the details of which we omit for the sake of brevity.

The basis concepts of kinematics are the structures of particle systems and their paths in space. To introduce the structure of a particle system we define a term $s_1$ by the type characterization:

$(T_1)$ $\quad s_1 \in P(\mathcal{J})$

and the relation

$(R_1)$ $\quad |s_1| \in \mathbb{N}$.

Let $\sigma_1$ be the set of all particle systems on $\mathcal{J}$. If $\pi \in \sigma_1$, we call $|\pi|$ the number of particles of $\pi$. In general, a further structure is introduced to separate different sorts of particle systems - different with respect to the effects of various forces (comp. [9], p. 77). But in our case, there is at the best only one sort of force and consequently we may dispense with a further characterization.

Paths of particles of a system are determined by a structure $s_2$, defined by

$(T_2)$ $\quad s_2 \in P(\mathcal{J} \times \Theta \times X)$

$(R_2)$ $\quad$ there is a $\pi \in \sigma_1$ such that $s_2$ is the graph of the function $r_\pi : \pi \times \Theta \to X$ and $r_\pi(i,t)$ is twice continuously differentiable with respect to $t \in \Theta$.

Let $\sigma_2$ be the set of all structures $s_2$ which satisfy $T_2$ and $R_2$, i.e. the set of graphs of all possible orbits.

Now, let us turn to Newton's theory, in particular to the Newtonian dynamics. To describe gravitational paths, we must first introduce the concept of mass. For this purpose we define a further structure $s_3$ by the typication

$(T_3^N)$ $\quad s_3 \in P(\mathcal{J} \times \mathbb{R})$

and the relation

$(R_3^N)$ $\quad$ there is a $\pi \in \sigma_1$ such that $s_3$ is the graph of a function $m_\pi : \pi \to \mathbb{R}$ and $m_\pi(i) > 0$ for all $i \in \pi$.

Again, let $\sigma_3^N$ be the set of all structures (mass functions) which satisfy $T_3^N$ and $R_3^N$. $m_\pi(i)$ is said to be the mass of the particle $i$ in the system $\pi$. After these preparations, Newton's law of gravitations reads as usual

$$(A^N) \quad \ddot{r}_\pi(i,t) = - \sum_{\substack{j\in\pi \\ j\neq i}} m_\pi(j) \frac{r_\pi(i,t) - r_\pi(j,t)}{|r_\pi(i,t) - r_\pi(j,t)|^3}, \; i \in \pi.$$

whereby we omit the constant of gravitation. Clearly, every tuple $(\pi, r_\pi, m_\pi)$ which satisfies relation $A^N$ is a model of Newton's theory.

In analogy with the mass function we define a function restating Kepler's constant for planetary systems. We introduce a structure $s_3'$ by the axioms

$(T_3^K)$ $\quad s'_3 \in P(J \times \mathbb{R})$

$(R_3^K)$ $\quad$ there is a $\pi \in \sigma_1$ and a $k_o \in \pi$ such that $s_3'$ is the graph of a function $\mu_\pi^{k_o}: \pi \to \mathbb{R}$, $\mu_\pi^{k_o}(i) = \delta_{k_o i}\mu_{k_o}$ and $\mu_{k_o} > 0$.

Here, $\delta_{ki}$ is the Kronecker symbol. Let $\sigma_3^K$ be the set of structures $s_3'$, and let us call $\mu_\pi^{k_o}$ the Keplerian function of the system $\pi$; the particle $k_o \in \pi$ is said to be the "sun" of system. This characterization of $\mu_\pi^{k_o}$ will be obvious from the following system of differential equations which restates the three Keplerian laws (comp. [13], p. 104):

$$\ddot{r}_\pi(i,t) = - \sum_{\substack{j\in\pi \\ j\neq i}} \mu_\pi^{k_o}(j) \frac{r_\pi(i,t) - r_\pi(j,t)}{|r_\pi(i,t) - r_\pi(j,t)|^3}, \; i \in \pi;$$

$$\frac{1}{2}|\dot{r}_\pi(i,t) - \dot{r}_\pi(k_o,t)|^2 - \mu_{k_o}|r_\pi(i,t) - r_\pi(k_o,t)|^{-1} < 0,$$

$$i \in \pi, \; i \neq k_o.$$

The first set of equations determines the motion of Keplerian systems by a "quasi-dynamic" with respect to the function $\mu_\pi^{k_o}$ i.e. there is no interaction between the planets $i \in \pi (i \neq k_o)$, but only a (Newtonian) interaction between the sun $k_o$ and the planets. The second set of equations "causes" the particles to move on closed orbits (ellipses). Observe that the orbit of the sun is an inertial curve.

Let us return to the general concepts of theory approximation. First we have to introduce some uniformities on the set of structures. This will be done in greater detail in the next chapter by the use of empirical imprecision-sets. For our purposes here, we shall assume that both theories are equipped with uniformities. Then, the physicists statement - "the theory T is approached by the theory T' "or" T is somewhere around T'" - makes sense, if the imprecision-sets of the uniformities are used to make this proximity relation precise. In my opinion, this was first done by Ludwig in his relation of blurred embedding[5] ([9] § 8, p. 98 ff). Roughly, T and T' satisfy that relation, if there are imprecision-sets in both uniformities such that the models of T' are embedded in the blurred set of models of T (blurred by the imprecision-sets of T) and the complement of the blurred set of T'-models (blurred by the imprecision-sets of T') is embedded in the complement of the set of T-models. In the simple case "Kepler-Newton", Ludwig's relation is fulfilled, if a certain subset of Newtonian models is included in the blurred set of Keplerian models and if the set of Keplerian models is included in the blurred subset of Newtonian models[6]. Moreover, there are no bounds for the admissible imprecision-sets in general. This is just the case which we have called a (physical) proximation of theories.

Ludwig regards the embedding relation as a rough introduction to the complicated relationships of physical theories (comp. [9], p. 99). This will be clear if one observes that the blurred embedding exists trivially for sufficiently large inaccuracies, but it does not generally exist for very small inaccuracies. There is a further argument which emphasizes the preliminary status of theory proximation. Many examples of theories which are in a limit-case relationship, to each other are made mathematically precise by exact convergence structures. It is impossible to refer comprehensively to the voluminous literature on this subject[7]; some examples may be found in [1] and [4] for the case "geometrical optics-electrodynamics", [6] and references cites there for the case "Newton-Einstein", [12] representatively for the case "thermodynamics-statistical physics" (comp. also the contribution of R. Werner in this volume) and [7] for the case "classical mechanics-quantum theory". In short, physical proximation is a first characterization of a more elaborate relation which is usual known as approximation of theories. The above mentioned examples illustrate that it is the so-called reducing theory T' which approximates the reduced theory T and not vice versa. Moreover, approximation in T' "generates" some structures (e.g. functions) which satisfy the fundamental laws of the reduced theory, i.e. structures isomorphic to the structures of T. But isomorphism is obtained only at the "end" of the limiting-processes and thus, we have to deal mainly with some suitable "limit-elements". In general, however, these limit-elements are not elements of the sets in which the approximation was performed. And here, the wheel comes full circle. It is just the uniformity which allows a concept of convergence without knowing the limit-elements and which generates all limit-elements by the operation of completion.

In this connection, we understand our approach of theory approximation (comp. [10]) as a first attempt to specify the relation of blurred embedding, or more generally, to make more precise the relationship of proximation and approximation. It is said that T is a limit-case of T', if T' is completed (with respect to an appropriate uniformity) in such a way that the completed theory contains structures isomorphic to the reduced theory T. In general, the concept of an isomorphism may be restated by an exact reduction relation, for instance by Ludwig's relation of (exact) embedding or standard embedding (comp. [9], p. 88, 90).

For abstract notions, let $\mathcal{U}_i'$ be a uniformity on the set $\sigma_i'$ of structures of the species $\Sigma_i'$ in the reducing theory, i = 1,...,m.

$$(B_1',\dots,B_n';\ \sigma_1',\dots,\sigma_m';\ A')_{\mathcal{U}_1',\dots,\mathcal{U}_m'}$$

By the operation of completion with respect to $\mathcal{U}_i'$ (comp. [2] II, § 3.7) we obtain the spaces $\hat{\sigma}_i'$ and in an analogous way the set $\hat{A}'$ as subspace of the completed product space $\widehat{\prod_i^m \sigma_i'}$ (comp. [2] II, § 3.9). Consequently, a completion $\hat{T}'$ of the theory T' is defined by the tupel

$$(B_1',\dots,B_n';\ \hat{\sigma}_1',\dots,\hat{\sigma}_m';\ \hat{A}')_{\hat{\mathcal{U}}_1',\dots,\hat{\mathcal{U}}_m'}$$

Suppose further an exact reduction relation R(T,T') reduces T to T', then we define an approximative reduction of theories:

Definition: The theory T is a limit-case of the theory T', AR(T,T'), in short, AR reduces approximatively T to T' iff there are separating uniformities in T' such that the completion $\hat{T}'$ satisfies $R(T,\hat{T}')$.

Observe that AR agrees with the relation R, if the reducing theory is already complete with respect to the presupposed uniformities. Thus, in accordance with the usual ideas of theory reduction, R is a special case of AR.

By definition, approximative reduction is mainly concerned with the uniform structures on which the completion is established. They have only to ensure the existence of structures which are (in some sense) isomorphic to corresponding structures of the reduced theory. Of course, this is a rather abstract characterization of the uniformities in T' which accomplish the approximation. For more insight into the complicated relationships of theory reduction, one has to examine the empirical significance of the uniformities.

## III. EMPIRICAL UNIFORMITIES AND CANONICAL UNIFORMIZATION

Empirical uniformities[8] are defined by imprecision-sets which characterize real measurements (comp. [9], § 6). They represent the fact that every measurement yields a result within an imprecision-interval. In general, physical measurements concern those entities which correspond (Abbildungsprinzipien) to elements of the base sets of a theory. Therefore it may be convenient to introduce empirical uniformities $\mathcal{V}_1,\ldots,\mathcal{V}_n$ on the base sets $B_1,\ldots,B_n$ ). We have already mentioned that it is characteristic of these uniformities that they are not uniquely determined by measurements. Only some topological properties are suggested, for instance the Hausdorff-property, local- or pre-compactness, and metrizability (comp. [9] § 9). Nevertheless, we remark that the choice of an empirical uniformity, or more precisely, the choice of a certain (admissible) subset of imprecision-sets is an important criterion with respect to the validity of a physical theory. In history there are many examples of physical models which turn into anomalies by transition to other (finer) imprecision-sets.
By canonical uniformization we understand a natural transport of the empirical uniformities from the base sets to the sets of structures. This may be accomplished by the following two basic operations.

First, we define a uniformity on a product set, the so-called product uniformity. Let X and Y be two uniform spaces with corresponding uniformities $\mathcal{U}_X$ and $\mathcal{U}_Y$; $U\in\mathcal{U}_X$ and $V\in\mathcal{U}_Y$.

$$\mathcal{W}(U,V):=\{(x,y;x',y')\in(X\times Y)\times(X\times Y)\mid(x,x')\in U,(y,y')\in V\}.$$

As U,V are running through the uniformities $\mathcal{U}_X$, $\mathcal{U}_Y$ the entourages $\mathcal{W}(U,V)$ form a fundamental system of a uniformity on X×Y, the product uniformity $\mathcal{U}_{X\times Y}$ (comp. [2] II, § 2.6). A similar definition holds for a family $(X_i)_{i\in J}$ of uniform spaces.

The second operation states the transport of the uniformity from the set X to its power-set P(X). For this purpose, we define the U-neighborhood of a subset M⊂X, $U\in\mathcal{U}_X$

$$U(M):=\{x\in X\mid \vee x'\in M\ :\ (x,x')\in U\}.$$

Consequently, we call two subsets M,N⊂X U-close, if each one is contained in the U-neighborhood of the other. The concept of U-closeness suggests the following definition.

$$\mathcal{W}(U):=\{(M,N)\in P(X)\times P(X)\mid M\subset U(N)\wedge N\subset U(M)\}.$$

As U runs through the uniformity $\mathcal{U}_X$ the entourages $\mathcal{W}(U)$ form a fundamental system of a uniformity $\tilde{\mathcal{U}}_X$ on P(X) (comp. [2] II, § 1,

Ex. 5). In contrast to the product uniformity, we observe that the Hausdorff-property of X is not transported to P(X). We shall avoid unessential complications in the following, if we work only with the so-called associated uniformity of $\tilde{\mathcal{U}}_X$. It is characterized by the equivalence relation

$$M \sim N \text{ iff } (M,N) \in \tilde{U} \text{ for all } \tilde{U} \in \tilde{\mathcal{U}}_X$$

on P(X). After transition to the uniform space $P(X)|_\sim$ the Hausdorff-property is obtained. Equivalently, we can uniformize the set $\mathcal{F}(X)$ of closed subsets of X, and $P(X)|_\sim$ is isomorphic to the uniform space $\mathcal{F}(X)$.

We will illustrate these operations on the set $\sigma_2$, the set of graphs which describe the motion of particle systems. We presuppose the empirical uniformities $\mathcal{V}_y, \mathcal{V}_\Theta$ and $\mathcal{V}_x$ on the base sets J, Θ and X. Let $\tilde{\mathcal{U}}_2$ be the canonically transported uniformity on $P(J \times \Theta \times X)$. Any element $g \in \sigma_2$ is a set of triples (i,t,x) i.e. an element of $P(J \times \Theta \times X)$. The relation, "two elements $g,g' \in \sigma_2$ are $\tilde{U}$-closed", i.e. $(g,g') \in \tilde{U}, \tilde{U} \in \tilde{\mathcal{U}}_2$, is equivalent to the following relations:

$$\underset{\text{def}}{\Leftrightarrow} \quad g \subset U(g') \wedge g' \subset U(g)$$

$$\Leftrightarrow \quad \bigwedge(i,t,x) \in g \bigvee(i',t',x') \in g' : (i,t,x;i',t',x') \in U$$

and vice versa

$$\Leftrightarrow \quad \bigwedge(i,t,x) \in g \bigvee(i',t',x') \in g' : (i,i') \in V_J \wedge (t,t') \in V_\Theta \wedge (x,x') \in V_X$$

and vice versa.

where $U = \mathcal{W}(V_J, V_\Theta, V_X)$ with $V_J \in \mathcal{V}_J$, $V_\Theta \in \mathcal{V}_\Theta$ and $V_X \in \mathcal{V}_X$.

Obviously, g and g' are $\tilde{U}$-close iff the particles, time and space points are V-close, ( V stands for the various generating entourages of $\tilde{U}$); in short, g is contained in the "imprecision-tube" of g' and vice versa.

For the sake of simplicity, let us restrict our consideration to the subset $\sigma_2(\pi) \subset \sigma_2$, i.e. to all possible orbits of a fixed particle system $\pi$. If we now adopt the discrete uniformity on the base set Θ - for instance, we are only interested in spacelike inaccuracies - we obtain a uniformity $\tilde{\mathcal{U}}_{2,\Theta}$ on $\sigma_2(\pi)$ which is called the uniformity of uniform convergence (comp. [2] X, § 1.1). Obviously, this uniformity only depends on the empirical uniformity $\mathcal{V}_X$, and $\tilde{\mathcal{U}}_{2,\Theta}$ is finer than the uniformity $\tilde{\mathcal{U}}_2$. By this construction we enter a new class of uniformities, the so-called uniformities of $S$-convergence (comp. [2] X, § 1.2). To obtain the variations of this type of convergence we introduce a set $S$ of subsets of $\Theta$. Then we define for $\theta \in \Theta$ and $g \in \sigma_2(\pi)$

$$g_\theta := \{(i,t,x) \in g \mid t \in \theta\}.$$

Clearly, $g_\theta$ is a certain segment of g which corresponds to the position function of $\pi$, restricted to times of $\theta$. By the use of $S$ we get a new entourage

$$\tilde{\mathcal{W}}(\theta,\tilde{U}) := \{(g,g') \in \sigma_2(\pi) \times \sigma_2(\pi) \mid (g_\theta, g'_\theta) \in \tilde{U}\}.$$

As $\theta$ runs through $S$ and $\tilde{U}$ runs through $\tilde{\mathcal{U}}_2$, the finite intersections of the entourages $\tilde{\mathcal{W}}(\theta,\tilde{U})$ form a fundamental system of a uniformity $\tilde{\mathcal{U}}_2^S$ which is coarser than $\tilde{\mathcal{U}}_2$. Again, let $\Theta$ be endowed with the discrete uniformity, then $\tilde{\mathcal{U}}^S_{2,\Theta}$ is said to be the uniformity of $S$-convergence, and it is coarser than $\tilde{\mathcal{U}}_{2,\Theta}$. For instance, if $S$ is the set of compact subsets of $\Theta$ (with respect to a given topology or uniformity), then g and g' are $\tilde{U}$-closed only with respect to a compact interval of time (comp. [2] X, § 1.3). It is just this type of uniformity which fits empirical methods of measurements better than finer uniformities[10].

IV. THE LIMIT-CASE EXAMPLE: KEPLER-NEWTON

In this chapter we will prove our main theorem - the limit-case relation between Kepler's theory of planetary motion and Newton's theory of gravitation. The argument is accomplished in two steps. First, we demonstrate that any n-particle model of Kepler's theory is approximated by a sequence of n-particle Newtonian models. The approximation succeeds in the compact convergence structure, generated by the usual (euclidian) uniformity of $\mathbb{R}^3$. After that we show the approximative reduction with some empirical uniformities. Here, the common kinematics implies a trivial embedding.

Before beginning we introduce some more mathematical structures. To formulate the usual ordinary differential equations we suppose that the base sets $\Theta$ and X are endowed with structures isomorphic to the usual structures on $\mathbb{R}$ and $\mathbb{R}^3$ respectively. For the present we restrict our consideration to an n-particle system and omit the index $\pi$ of a system. Therefore, we define by

$$\langle n\rangle := \{1,\ldots,n\}$$

$$r : \langle n\rangle \times \mathbb{R} \to \mathbb{R}^3$$

$$m : \langle n\rangle \to (0,\infty)$$

a position and mass function of the system $<n>$. Finally, we will characterize by $r_K$ and $r_N$ a Keplerian and Newtonian position function respectively, i.e. solutions of the corresponding differential equations (comp. chapt. II).

Theorem 1: For every $\varepsilon > 0$ and for every Keplerian orbit $r_K$ and every compact interval $\theta$ of $\mathbb{R}$ there is a mass function with a Newtonian orbit $r_N$ such that

$$|r_N(i,t) - r_K(i,t)| < \varepsilon$$

for all $i \in <n>$ and $t \in \theta$.

Proof: Without loss of generality let $k_o = 1$ be the Keplerian sun. We define:

$$(X_{3i-2}(t),\ X_{3i-1}(t),\ X_{3i}(t)) := r_N(i,t)$$

$$(X_{3(n+i)-2}(t),\ X_{3(n+i)-1}(t),\ X_{3(n+i)}(t)) := \dot{r}_N(i,t)$$

$$m_i := m(i),\ m := (m_2,\dots,m_n) \quad \text{for all } i \in <n>.$$

Using the Newtonian equations we obtain a system

$$\text{(E)} \qquad \dot{x} = f(x,m)$$

Of differential equations with $x := (x_1,\dots,x_{6n})^t$ and $f(x,m) := (f_1(x,m),\dots,f_{6n}(x,m))^t$. If $m = 0$ and $m_1 = \mu_1$, the system E coincides with the first set of Keplerian equations. Obviously, $f_i$ is continuously differentiable with respect to $x_j$ and $m_k$ (for all admissible $j$ and $k$) on the set $G := \mathbb{R}^{7n-1} \setminus C \times \mathbb{R}_+^{n-1}$, where C is the set of collision points $(x_1,\dots,x_{6n})$ with $r_N(l,t) = r_N(l',t)$ for $l \neq l'$. Consequently, f satisfies a Lipschitz condition with respect $(x,m)$ on G. Thus, by a theorem about the dependence of solutions on parameters (comp. [8] III.4), the solution $x(t;x_o,m)$ of E is a Lipschitz continuous function on its open domain of definition

$$D := \{(t,x_o,m)\,|\,(x_o,m) \in G \wedge t \in J(x_o,m)\}$$

where $J(x_o,m)$ is the maximal interval of the solution $x(t;x_o,m)$ $(x_o := x(t_o))$. Moreover, the solution is bounded on every compact subset C of D such that x satisfies a Lipschitz condition on C. By the way, x is arbitrarily often differentiable on D with respect to the components of $x_o$ and m. Finally, for a given Keplerian initial value $(x_o,m)$, $\varepsilon > 0$ and compact $\theta \subset \mathbb{R}$ we obtain

$$|x(t;x_o,m) - x(t;x_o,0)| \leq \text{const.}|m|$$

for all $t \in \theta$. By choosing $m \leq \frac{\varepsilon}{\text{const.} + 1}$, the theorem is proved.

From this theorem we may conclude that any Keplerian model is approximated by a sequence of Newtonian models with respect to the uniformity of compact convergence. The sequence corresponds uniquely to a minimal Cauchy filter on the uniform space of Newtonian models which just approximates the given Keplerian model. Thus, in the sense of the Hausdorff completion, the minimal Cauchy filter is an element of the completed set and moreover, the element is to be identified with the Keplerian model. Consequently, the completion includes the set of Kepler models.

These more informal statements have to be made more precise in our formal framework. Let us assume some empirical uniformities $\mathcal{V}_J$, $\mathcal{V}_\Theta$, $\mathcal{V}_X$ on the base sets $J$, $\Theta$, X. In particular, the last two uniformities are presupposed to be not finer than the usual uniformities on $\mathbb{R}$ and $\mathbb{R}^3$. We suppose that this condition is not a serious restriction to the concept of the empirical uniformity. Corresponding to the empirical conception that any two classical particles are distinguisable, the uniformity $\mathcal{V}_J$ may be the discrete uniformity on $J$. $\mathcal{V}_J$ is complete (comp. [2] II, § 3.3) and metrizable (discrete metric). Consequently, the canonically uniformized space $P(J)$ is complete (comp. a corresponding theorem in the contribution of H.-J. Schmidt in this volume or the hypercomplete spaces (comp. [5]). Finally $\sigma_1$ is closed in $P(J)$ and hence complete.

Now, let us turn to the remaining set of structures. For the present we shall restrict our consideration to one fixed particle system $\pi$. We presuppose the discrete uniformity on $\Theta$ and the usual uniformity on X (which is isomorphic to $\mathbb{R}^3$). Let $\mathcal{C}$ be the set of all compact subsets of $\Theta$, compact in the usual topology of $\mathbb{R}$. As we have shown in the last chapter we obtain the uniformity $\mathcal{U}^{\mathcal{C}}_{2,\Theta}$ on $P(J \times \Theta \times X)$. The subspace $\sigma_2(\pi)$ is isomorphic to the subspace $C^2_c(\langle n\rangle \times \mathbb{R}, \mathbb{R}^3)$ of $C_c(\langle n\rangle \times \mathbb{R}, \mathbb{R}^3)$, the set of continuous functions endowed with uniform structure of compact convergence. Clearly, $C^2_c$ is not closed in $C_c$ and hence not complete. We get the space $\sigma^N_3$ by the canonical uniformization of the set $P(J \times \mathbb{R})$. $\sigma^N_3(\pi)$ is isomorphic to the space $\mathcal{F}_{\mathcal{U}}(\langle n\rangle, \mathbb{R}_+)$ which is determined by the uniformity of uniform convergence (or equivalently by compact or pointwise convergence, comp. [2] X, § 1.3). Again, the space $\sigma^N_3(\pi)$ is not complete, because any Keplerian function is approximated by a Cauchy sequence of mass functions. Thus, all Keplerian functions are included in the completion $\hat{\sigma}^N_3$.

Finally, we consider the completion $\widehat{\sigma_1 \times \sigma_2 \times \sigma_3^N}$ which is isomorphic to $\sigma_1 \times \hat{\sigma}_2 \times \widehat{\sigma_3^N}$ ([2] II, § 3.9). We observe that the completed set $\widehat{A^N}(\pi)$ is the set of minimal Cauchy filters, which is divided in two classes: the set of filters which converge in $A^N(\pi)$ (Newtonian models) and the set of filters which do not converge. In the last set, the Keplerian models are included in virtue of theorem 1.
By the uniform structure of $\sigma_1$ we may generalize our argument for all particle systems, and we obtain

$$A^K \subset \widehat{A^N}$$

Now, we use the fact that any Cauchy filter with respect to a uniformity $\mathcal{U}$ preserves the Cauchy property under transition to any coarser uniformity $\mathcal{U}' \subset \mathcal{U}$. Recalling the Hausdorff property of the uniformities, we conclude that that relation $A^K \subset \widehat{A^N}$ holds for the empirical uniformities. Thus, we have proved our main result:

Theorem: Kepler's theory of planetary motion is a limit case of Newton's theory of gravitation.

## V. CONCLUSION

We have pointed out what kind of approximation satisfies the limit-case relationship between Kepler's and Newton's theory. The transition from gravitational dynamics to Kepler's planetary quasi-dynamics is mainly carried out in two steps: The planet-planet interaction made small by letting the masses become small and secondly, a sufficiently coarse structure had to be chosen to overcome long-time influences (secular perturbations). On the other hand the empirical informities set bounds to the coarseness. Within this frame the compact convergence turns out to be the appropriate structure. Finally, the restriction to finite intervals of time tallies well with the methods of measurement.

As a consequence, we can understand in which applications, e.g. in which representations of real particle systems, a replacement of the complicated gravitational equations by the simpler Keplerian equations makes sense. And we can estimate the differences which are produced by the replacement. In short, it is possible to characterize those domains of applications in which both theories are roughly equivalent and in which the gravitational description fits more with the empirical facts. In our opinion, this demonstrates the comparability and the progress which take place in theory dynamics.

NOTES

1) The term "imprecision-set" in chosen for the sake of accordance with other articles in this volume. It is intended to denote the same as "inaccuracy-set" used in my former publication [10].

2) Private communication of G. Süßmann.

3) The model is due to W. Hoering.

4) The sets $\sigma_i$ of structures of the species $\Sigma_i$ are called factor sets in [10]. Their product set $\sigma = \sigma_1 x \ldots x \sigma_n$ is called a theory matrix and its elements are comparable with the possible models in the theory concept of the Non-Statement view (comp. [14]).

5) The blurring of a relation is defined in [9] § 6, p.54. It is comparable with "logic of approximation by blurring" presented in [11] (comp. also Moulines' contribution in this volume).

6) Roughly, this is also the intention of Scheibe's approach (comp. [13] , p. 113). But Scheibe uses the smallest imprecision - the "blurring" is performed by the topological closure operation. Consequently, the first part of Ludwig's embedding relation is not satisfied.

7) Almost all text books of theoretical physics contain more or less detailed examples of limit-case relationships.

8) We use the term "empirical" in a quite different sense as Moulines in his contribution to this volume. Our empirical uniformities are defined on the base sets and are canonically transported to the set of structures (which correspond to the set $M_p$ of possible models). The empirical uniformity of Moulines is defined on $M_p$ and does not imply uniformities on the base sets in general. In particular, Moulines' condition D5(2) (pseudo-diagonal) does not satisfy our presupposed Hausdorff-property (comp. [10] , note 2).

9) In this chapter we shall omit the primes designating the reducing theory as there is no danger of confusion between the reduced and the reducing theory.

10) We recall our counter-example of approximation, the planetary roundabout model. Obviously, it is the uniformity of uniform convergence, generated by the usual uniformity which prevents the approximation of these models. But we shall prove an approximation with respect to the uniformity of compact convergence.

LITERATURE

[ 1] M. Born, Optik, Springer; Berlin 1933

[ 2] N. Bourbaki, General Topology;Herman, Paris 1966

[ 3] N. Bourbaki, Theory of Sets; Herman, Paris 1968

[ 4] R. Breuer, J. Ehlers, Propagation of high-frequency electromagnetic waves through a magnetized plasma in curved space-time I, II; Proc. R. Soc. Lond. A 370 (1980)

[ 5] E. Cech, Topological spaces; Interscience, London 1966

[ 6] J. Ehlers, Über den Newtonschen Grenzwert der Einsteinschen Gravitationstheorie, to appear in Festschrift für Mittelstaedt (1981)

[ 7] K. Hepp, The classical limit for quantum mechanical correlation functions; Commun. math. Phys. 35, 265-277 (1974)

[ 8] H.W. Knobloch, F. Kappel, Gewöhnliche Differentialgleichungen; Teubner, Stuttgart 1974

[ 9] G. Ludwig, Die Grundstrukturen einer physikalischen Theorie, Springer, Berlin 1978

[10] D. Mayr, Investigations of the concept of reduction II; Erkenntnis 16 (1981)

[11] C.-U. Moulines, Approximative explanation of empirical theories: a general explication; Erkenntnis 10, 201-229 (1976)

[12] D. Ruelle, Statistical Mechanics, Rigorous Results; Benjamin, New York 1969

[13] E. Scheibe, Die Erklärung der Keplerschen Gesetze durch Newtons Gravitationsgleichung; in E. Scheibe und G. Süßmann (eds), Einheit und Vielheit, Festschrift für Carl Friedrich von Weizsäcker, Vandenhoeck & Ruprecht, Göttingen, 1973

[14] W. Stegmüller, The Structuralist View of Theories; Springer, Berlin 1979

# G. LUDWIG'S POSITIVISTIC RECONSTRUCTION OF THE PHYSICAL WORLD AND HIS REJECTION OF THEORETICAL CONCEPTS

A. Kamlah
Fachbereich 2 der Universität Osnabrück
Postfach 44 69
4500 Osnabrück
Federal Republic of Germany

## I. INTRODUCTION

In this paper we investigate some aspects of G. Ludwigs rational reconstruction of physical theories and try to compare his ideas with conceptions of analytical philosophers of science. We shall find striking similarities between Ludwigs philosophy of physics and the neopositivist account of the Vienna Circle before world war II. Ludwig started his philosophical investigation of physics quite independently of the analytical philosophical tradition. He was led to his considerations when he studied the foundations of quantum mechanics and felt a need for a general analysis of physical theories. He used the logic and set theory of Bourbaki as his tool and so his logical terminology differs from that used by analytical philosophers of science. His philosophical vocabulary moreover is completely different from any language ever used by a philosopher. Even quite common words like "hypothesis", "possible" or "wrong" have nonstanstard meanings in his writing. Therefore his books and papers are not easily understood by the average philosopher of science. Before entering into an examination of his claims, one has to learn Ludwig's language.

In spite of these difficulties Ludwig's writings seem to me to be interesting enough to merit serious investigation. We want to give here a translation of some of his ideas and compare them with current issues of philosophy of science. Furthermore we want to discuss and criticize Ludwig's claim that physics may be rationally reconstructed without application of theoretical terms.

## II. BASIC STATEMENTS AND CORRESPONDENCE RULES

A physical theory PT according to Ludwig consists of a mathematical theory MT, the so called domain of "reality" (Wirklichkeitsbereich") W, and the correspondence rules ("Abbildungsprinzipien") (—). (We shall motivate our translation for "Abbildungsprinzipien" when we discuss them in detail.) Thus a physical theory PT is a triple MT(—)W. The mathematical theory is just what a mathematician would expect it to be, if he is familiar with the modern set theoretical treatment of axiom systems, and so I can leave it out of consideration here.

It is more difficult however to understand Ludwig's domain of "reality" and his correspondence rules. We shall start by investigating the former, and in order to do this we shall first try to understand the "basic domain" ("Grundbereich"), which is the most important part of the reality domain. When this has been done, it will be easy to make a transition from the part to the whole. The basic domain is the totality of facts, which are empirically relevant for the theory PT, thus the basic domain is what usually is called the empirical basis of the theory. This basis however does not consist of propositions. Ludwig says that it is rather a set of physical processes than a set of propositions describing them (1974, p. 61). Fragments of the basic domain are called by Ludwig "Realtexte" (collections of facts, see Hartkämper and Schmidt in this volume). A physicist is never occupied with all empirical facts, which are important for a theory. His empirical material cannot contain more facts, than he is able to record. Therefore what is in fact given to him is not the whole basic domain but rather a certain "Realtext". Ludwig's conception of the experimental material as a text reminds us of the old allegory that the physicist is reading in the book of nature. The "Realtext" is "normed" and "read" by the physicist. Norming the "Realtext" means naming the physical entities appearing in it, for instance space-time points, physical bodies, and particles. Reading means here expressing the relevant facts in propositions, which have been called "basic statements" by K.R. Popper. Ludwig points out that we need "pre-theories" ("Vortheorien") in order to read the "Realtext". These are those theories which have to be established before the theory PT can be tested, which are needed to understand the experimental apparatus. Thus mechanics is a pre-theory for electrodynamics; we have to be able to talk about forces before we can measure the strength of an electrical field. Electrodynamics is a pre-theory for nuclear physics; we have to be able to talk about electrical fields before we make experiments with betatrons or cyclotrons. The language of these pre-theories is in fact the observational language used for formulation of basic statements. We have to apply, what C.G. Hempel calls the "antecendently available vocabulary" for the physical theory, an observational language, which is sufficiently sophisticated to describe the experiments by which PT is tested, but which is still lacking the specific concepts of the theory PT itself. Hempel writes:

> "The phenomena which a theory is to explain as well as those by reference to which it is tested are usually described in terms which are by no means observational in a narrow intuitive sense, but which have a well-established use in science and are employed by investigators in the field with high intersubjective agreement. I shall say that such terms belong to the <u>antecedently available vocabulary</u>. Often, such terms will have been introduced into the language of science in the context of an earlier theory. For example, Bohr's and Sommerfeld's theories of atomic structure were developped to account for certain characteristic features of the spectra of chemical elements. Those features were described in terms of wavelength and intensities of radiation emitted or absorbed, and thus by means of a vocabulary that clearly is not observational in an intuitive sense; yet this vocabulary was used by physicists with high accuracy and interpersonal uniformity; the principles for their use, e.g., for the measurement of wavelengths, having been provided by earlier theories, including wave optics. It seems reasonable, therefore, to construe the interpretation base of a theory as consisting not of observational predicates, but of antecedently available ones.
>
> The concept of antecedent availability is, again, relational: a predicate, say 'electrically charged' or 'introverted', cannot be said to be antecedently available <u>tout court</u>, but only with respect to the introduction of a given theory." (1973, p. 372-373.)

Though Ludwig does not talk of the language used for reading the "Realtext", it seems to be quite clear, that for him the facts of the basic domain have to be described by an antecedently available vocabulary in the sense of Hempel. He says in his most recent book: "... the stated facts can be directly or indirectly formulated in language with the aid of pre-theories." (1978, p. 33. He is less clear about that matter in his earlier writings.)

So far we have only talked about the basic domain, which is only a part of the domain of reality described by the theory. When we have understood the basic domain however, we may easily grasp the concept of domain of reality. The domain of reality is simply the totality of facts which may be concluded from the basic statements and the theory. These include predictions and retrodictions made by the theory, i.e. facts of the same type as those of the basic domain, which may be described by the antecedently available vocabulary, "directly observable facts" (1978, p. 184), and in addition "indirectly observable facts", which are described by propositions about physical entities, hypothesized by the theory PT. Thus the concept of the domain of reality does not lead to any difficulty. However, another question is, why Ludwig uses the expression "Wirklichkeitsbereich" (domain of reality) for denoting the set of inferred facts. Does he believe that this is all the theory is about? Suppose we are dealing with a theory about electrons. Most of them have never been involved in any observation. Only those electrons, from which some data may be inferred from observable facts are objects in the reality domain. Does Ludwig want to say by coining his term "Wirklichkeitsbereich" that the other electrons do not exist for the theory? But the reader of Ludwig's writings should not ask for the motivation of his terminology. He would be left with too many unsolved problems.

We turn now to the second constituent of PT = MT(—)W, namely the correspondence rules (—). Ludwig calls them "Abbildungsprinzipien", which indicates that they map some of the terms of the antecedently available vocabulary into some terms of the language of the theory MT. A mathematical theory is in the first instance an uninterpreted calculus, the signs appearing in it have no meaning by themselves. If such a theory is to function as a physical theory, i.e. as an interpreted calculus, at least some of its signs have to be endowed with a meaning. The terms which are mapped onto corresponding terms of the antecedently available vocabulary are called by Ludwig "Bildrelationen" (interpreted relations) and "Bildterme" (interpreted terms). Since we do not want to apply here the Bourbaki distinction between terms and relations, we shall talk here only of picture <u>terms</u>, understanding "term" in the usual way as any simple or complex sign of a language, which is no sentence. Thus in many cases, in which relations à la Bourbaki are not sentences, they are terms in our sense. What has been said until now seems to indicate clearly that Ludwig's "Abbildungsprinzipien" are indeed nothing else than Carnaps correspondence rules. Therefore I shall designate them by the more familiar expression of Carnap, as was already done by H.W. Schürmann (1977, p. 24: "Zuordnungsregeln"). By the correspondence rules the external basic statements which are expressed in the antecedently available vocabulary may be translated into corresponding internal basic statements in the language of the mathematical theory. These internal basic statements are called by him "Abbildungsaxiome" (axioms of the observational report). By the choice of this expression Ludwig indicates that an internal basic statement is an ultimate premise, i.e. an axiom for the inference of the physicist, and that it mirrors or maps a fact belonging to a "Realtext". We may as well call them internal basic statements, if we remind ourselves that a basic statement in the language of a theory is something different from an external basic statement formulated in some sort of observational language. Popper does not make this external-internal distinction, since he does not talk much about languages. It is evident however that we have to draw it in principle, if we are using a two-language model for the logical reconstruction of a physical theory, even if the distinction will frequently be neglected in the following discussion, where this cannot lead to mistakes.

I shall finish now my brief comparism between Ludwig's account and Hempel's revised "standard construal" of the relation between facts and theories in physics. I have given here an interpretation of Ludwig's ideas which possibly is not correct and open to revision by Ludwig himself, who is the ultimate authority for any interpretation of his writings. Ludwig has not laid so much emphasis on the fact that the "Realtext" must be given in linguistic formulation before the correspondence rules are applied to it. In his earlier writings one is tempted to read Ludwig as if without correspondence rules one could not talk at all about physical facts, as if these rules connect terms of a language, not with signs of another language, but rather with the physical objects and facts themselves, and as if the "Realtext" is only "read" after the correspondence rules have been applied. Thus Ludwig would have understood the correspondence rules as a brigde between linguistic and nonlinguistic entities, between sign and objects, properties, and relations. It may be interesting to notice here that H. Reichenbach has the same tendency, namely to present his "coordinative definitions", which are very similar to Carnap's correspondence rules, as a link between nature and language. We are led here to the same difficulties just mentioned. (A. Kamlah 1977, p. 394-395.)

## III. LUDWIG'S REJECTIONS OF THEORETICAL TERMS

The sketch of Ludwig's physical semantics in the last section shows his approach to be very near to the standard construal, or more exactly, in good agreement with Hempels revised standard construal. In the introduction however I had claimed a striking similarity with the Vienna-Circle neopositivism, which now has to be demonstrated. It is Ludwig's rejection of theoretical terms and the thesis that physical concepts are reducible to those of the antecedently available vocabulary which Ludwig has in common with the neopositivists. Of course he does not express his claim in these terms. He would rather say that all concepts of a theory could be expressed by "intrinsic" terms of its "axiomatic basis". A theory is given in its axiomatic basis, if it is written down solely in purely set theoretical signs and in interpreted terms, which obtain their meaning directly from the correspondence rules. Thus one may say that interpreted terms are translatable into the antecedently available vocabulary by the correspondence rules and therefore that a theory in its axiomatic basis is also translatable into this observational language. An intrinsic term of a theory is - one can say it in this way - a simple or complex sign which may serve as a definiens for an explicit definition in this theory. ("intrinsic for a structure s" is a Bourbaki term (Bourbaki 1968, p. 266). We want to call a term "intrinsic in a theory" if it is intrinsic for all its models.) From these explanations of the concepts "intrinsic" and "axiomatic basis" it is quite clear that Ludwig defends the reducibility of theoretical concepts to the antecedently available vocabulary. In the following discussion the distinction between interpreted terms and concepts of the antecedently available vocabulary if often neglected, as has already been announced.

It is indeed puzzling that in our time a physicist is defending the definability of theoretical terms, while the proof of the contrary is generally considered as one of the most important achievements of modern philosophy of science. How shall we react? Shall we recommend to Ludwig some standard textbooks in philosophy of science and leave him alone with the task to correct his own errors? I do not think that such a reaction would be justified. Ludwig has arguments in favour of his position, and we should at least be ready to question or relativize the undefinability thesis, if these arguments are sound. This generally accepted thesis has probably been prematurely canonized, and a reconsideration might reveal some weak points in its defence. Therefore, I want to look for possible arguments for Ludwigs claim.

First possible argument: Definability and expressibility by an intrinsic term may be different properties of concepts after all. They are indeed, but definability is the wider of the two concepts, and intrinsic terms are equivalent to a subclass of definable terms. An intrinsic term is in any case an expression, which contains no other primitive terms than logical and set theoretical constants and the extralogical vocabulary of the theory. Thus one is inclined to say that expressibility by an intrinsic term is equivalent to explicit definability, which is stronger than unspecified definability. It is not as easy as it seems at first sight, however, to compare meta-concepts of ordinary and Bourbaki logic. The list of Bourbaki's logical constants contains the operator $\tau$ which cannot be interpreted in a truth functional semantic system without difficulties. Propositions like $B(\tau_x(A(x)))$ have in some cases no determinable truth value. Therefore one is either forced to say that the truth value of $B(\tau_x(A(x)))$ is unknown or that it is indeterminate. We are however not primarily interested in interpreting Bourbaki logic in terms of truth functional logic. We rather want to understand Ludwig's intention and to ask for the justification of his ideas, and therefore it is more important to us to ask how he understands intrinsic terms or their application in physical theories, than to know how these terms are defined by Bourbaki. Ludwig expects that the intrinsic terms which represent theoretical concepts like temperature, entropy, electric and magnetic fields, charge and current density, are given as functions of interpreted terms (or interpreted relations, 1978a, p. 104). It is, therefore, quite clear that Ludwig does claim the definability of theoretical terms in the sense of analytical philosophy of science.

Second possible argument for Ludwig's claim: Ludwig says in his book about fundamental structures of physical theories that the electrodynamical vector potential $(\vec{A},\varphi)$ is no "real" physical entity, even if the electromagnetic field $(\vec{E},\vec{B})$ is known (1978,p.179). There are always different vector potentials $(\vec{A},\varphi)$ belonging to one field $(\vec{E},\vec{B})$. By "real" Ludwig means something similar to "determinable form the given evidence". Even if this characterization may prove to be not quite correct under closer examination, it may be sufficient for the present purpose. The class of vector potentials $(\vec{A},\varphi)$ which are admissible for a known field $(\vec{E},\vec{B})$, however, is a "real" entity for Ludwig. It is also clear that the vector potential cannot be defined in an antecedently available vocabulary which describes the forces acted upon point charges by $\vec{E}$ and $\vec{B}$.

Nevertheless Ludwig holds the class of admissible vector potentials to be "real" if the electrodynamic field $(\vec{E},\vec{B})$ is known. If $(\vec{E},\vec{B})$ is considered to be definable in the antecedently available vocabulary by Ludwig, the class of admissible vector potentials might be definable too.

If we generalize this example, we might come to the conclusion that in cases where a concept is theoretical and where different extensions*) of the concept are compatible with the known extentions of the antecedently available vocabulary, Ludwig might say that these different extensions are not "real", the class of admissible extensions, however, is indeed "real". Thus Ludwig's claim of definability of theoretical concepts might show up to be in fact a more modest thesis merely claiming that the set of admissible extensions is a definable concept. So for example he might say that in a mechanical system the class of all sets of mass values which is compatible with the observed time dependent position functions for n particles may be defined in the language of kinematics.

Such an interpretation of Ludwig's thesis, however, would be a considerable trivialization. In a recent paper Ludwig himself has introduced the difference between "theoretical concepts in the customary sense (1981). A typical theoretical auxiliary concepts" ("theoretische Hilfsbegriffe") and theoretical auxiliary concept is the phase of a quantum mechanical wave function. These concepts do not characterize any measurable quantities. Ludwig says clearly that the typical examples for theoretical concepts in philosophy of physics do not belong to this category. Therefore we may suspect that Ludwig is defending the following thesis, which we give here in a short version:

*What can be measured can also be defined.*

We have therefore to look for other reasons for Ludwig's scandalous claim.

Third possible argument for Ludwig's claim: We are coming now to consideration which have to be taken much more serious than the preceding ones. Ludwig may apply an overtly or covertly non-exten-

*) The extension of a predicate is the set of objects or ordered sets of objects for which it holds or is true.

sional language. Propositions in an extensional language do not change their truth values if some expressions appearing in them are replaced by others having the same extensions. Therefore the "substitution rule" is valid for them. The thesis of non-definability of theoretical terms is defended by analytical philosophers of science for extensional languages. They do not claim anything for non-extensional languages, which violate the above substitution rule.All languages, which talk about possible worlds different from the real one are non-extensional.

A simple example may show this clearly. In all possible worlds 12 is an even number. Therefore P, the proposition "12 is necessarily even", is true. If P would belong to an extensional language, we might replace "12" by "the number of apostles", since in our real world there have been 12 apostles. And we could conclude that the number of apostles is necessarily even, which is not true, since in some possible worlds there might be an odd number of apostles. Thus we are led to the result that P is not a proposition in an extensional language. This conclusion may be generalized to the theorem that all languages which talk about possible worlds are non-extensional, and violate the substitution rule.

A language which allows for operational definitions of the following form is naturally also non-extensional: "The quantity R has for the object a at time t the value r, if by application of the operation O at time t to a we would obtain (or would have obtained) the numerical value r." An example may be: "The temperature measured in $^{o}C$ of a at time t is r, if we would obtain (or would have obtained) at time t by measurement with a mercury thermometer at a just the value r $^{o}C$." Definitions of this form make use of counterfactual conditionals and talk about possible worlds which are different from the real world. Therefore they do not obey the rules of an extensional logic.

If Ludwig would use a language which contains propositional operators for counterfactual conditionals, he could of course define at least some theoretical terms. His logic, the Bourbaki logic, however is extensional, and thus he has no modal logic or logic of conditionals at his disposal. Therefore Ludwig's thesis cannot be substantiated by application of such a logic. If we want to explain, why Ludwig defends a generally rejected claim by his possible application of non-extensional languages, we have to refer to languages which talk in a covert way about non-real possible worlds. In physics we often find concepts which characterize the behaviour of physical systems as functions of their states. Such functions are the Lagrangian and Hamiltonian function, potential functions and force fields. These functions inserted into corresponding equations tell us, what would happen, if the physical systems were in certain states. So they inform us about these systems in non-real possible worlds. They are not nomological functions, since they are not completely determined by natural laws themselves, but in macroscopic physics at least they contain additional information about dispositional properties of the individual physical systems in question. In microphysics the Hamiltonian operator is completely nomological and therefore of a different nature. Functions of the just characterized mixed type I want to call "crypto-intensional functions", and those which describe only the real world, I want to call "purely extensional". This is a preliminary definition, which will be replaced by a better one later.

We should now study some physical examples, in order to understand thorougly how crypto-intensional functions make definitions of otherwise theoretical concepts possible. Let us imagine that we can measure forces and positions of particles without difficulties. Let the terms "force" and "position" belong to our antecedently available vocabulary, or to the terms of our pre-theories. Let this not be the case for electrical charge. So "eletrical charge" is a term of the theory we are concerned with, namely electrostatics or the theory of Coulomb forces, and we may suspect that it is a theoretical term relative to the given antecedently available vocabulary.

We shall start by discussing our example in terms of purely extensional functions. We shall first assume that forces $\vec{f}_i(t)$ and positions $\vec{s}_i(t)$ are vector functions of particle number and time. Let M be the set of n particles and T the time interval, we have the following typification for $\vec{f}$ and $\vec{s}$:

$$\vec{f} : M \times T \to \mathbb{R}^3 \; ; \quad \vec{s} : M \times T \to \mathbb{R}^3 \; .$$

Electrical charge may be given as a function of the particle number and time (a particle may change its charge in time):

$$e : M \times T \to \mathbb{R} \; .$$

Now we have to state the most important axiom of our theory:

$$f_i = \sum_j \frac{e_i e_j \vec{r}_{ij}}{|\vec{r}_{ij}|^3} \; .$$

By this axiom we obtain a system of equations which "defines" electrical charge e in terms of force and particle coordinates.

In many cases these equations will have unique solutions; but there are also cases where many solutions can be found. If all n particles of the system are situated on a straight line, we may consider $\vec{f}_i$ and $\vec{r}_{ij}$ as scalars and have to solve n equations $f_i = \sum_j \frac{e_i e_j}{r_{ij}^2}$. Since from the conservation of momentum we may derive $\sum_i f_i = 0$, the last equation is simply the sum of the other n-1 equations, and we have therefore to solve n-1 equations for n unknown quantities. Therefore there are at least some cases in which electrical charges cannot be determined by known forces and mutual distances of n particles, and thus we cannot consider electrical charges as being definable in electrostatics or in the theory of Coulomb forces.

The whole situation is different, if we consider force as a function of all particle coordinates. Let the force $\vec{F}$ be a function of the particle index of the particle affected, of time, and the position of all particles of the system:

$$\vec{F} : M \times T \times \mathbb{R}^{3n} \longrightarrow \mathbb{R}^3$$

Now F is a function of the configuration of the system, i.e. of the same system in all possible states or of possible worlds in which it exists. Thus $\vec{F}$ is a crypto-intensional function and no purely extensional function. In the new theory we have no difficulties to define electrical charge by Coulomb's law, and the forces and distances involved. There is an infinite number of configurations and many of these are not degenerated like the arrangement on a straight line, and will supply us with 3(n-1) equations for n unknown quantities. I shall also obtain a complete solution of the problem, if I look at pairs of particles at very large distances from the remaining n-2 particles, so that only the forces between two particles have to be taken into consideration. By this procedure I shall obtain the equations

$\vec{F}_{ij} = \frac{e_i e_j \vec{r}_{ij}}{|\vec{r}_{ij}|^3}$ or $e_i e_j = a_{ij}$, where $a_{ij}$ is given by

$a_{ij} = \vec{F}_{ij} \cdot \vec{r}_{ij} |\vec{r}_{ij}|$ and is an expression in terms of the antecedently available vocabulary. We may then simply define $e_i$ as

$|e_i| = \sqrt{\frac{a_{ij}\ a_{ik}}{a_{ik}}}$ with $\frac{e_i}{e_j} = \frac{a_{ik}}{a_{jk}}$ . The signs of the charges can only be determined up to a common factor ±1. Thus we see that by application of crypto-intensional functions we may define the electrical charge without difficulties, while this was not possible by the use of purely extensional functions.

We may add a second example in order to understand that there is a systematic connexion between the crypto-intensional versus purely extensional functions on the one side and the definibility versus undefinibility of theoretical concepts on the other. The example has been discussed by M. Heidelberger (1978) and concerns the definition of electrical resistance by Ohm's law. Again we shall apply first purely extensional and later crypto-intensional functions. Our antecedently available vocabulary consists of a set M of resistors, electrical current I as a function of the resistor index and time, and a set of electrical sources N. Thus current is typified as follows: (T being again a time intervall):
$I : M \times T \longrightarrow \mathbb{R}$ . Resistors and sources may be connected or not. We may express this by a three place relation con, con(i,k,t)

saying that resistor i is connected with source k at time t. The typification is for con: con $\subset M \times N \times T$.

By the theory two new concepts are introduced, voltage U and resistance R. Their typifications are:

$$U : N \rightarrow \mathbb{R}; \quad R : M \rightarrow \mathbb{R}.$$

The theory itself is rather simple. Its main axiom is:

$$i \in M \wedge k \in N \wedge t \in T \wedge \mathrm{con}(i,k,t) \rightarrow U(k) = R(i) \cdot I(i,t)$$

This is the well known Ohm's law. To this postulate we have to add two additional axioms which formulate two boundary conditions under which Ohm's law holds. A resistor can only be connected with one source and a source only with one resistor:

$$t \in T \wedge i \in M \wedge j \in M \wedge k \in N \wedge \mathrm{con}(i,k,t) \wedge \mathrm{con}(j,k,t) \rightarrow i = j$$

$$t \in T \wedge i \in M \wedge k \in N \wedge l \in N \wedge \mathrm{con}(i,k,t) \wedge \mathrm{con}(i,l,t) \rightarrow k = l$$

We may now define ratios of resistances by a "conditional definition":

$$t \in T \wedge t' \in T \wedge i \in M \wedge j \in N \wedge k \in N \wedge \mathrm{con}(i,k,t) \wedge \mathrm{con}(j,k,t')$$
$$\rightarrow \frac{R(i)}{R(j)} = \frac{I(i,t)}{I(j,t')} .$$

By the last "conditional definition" the ratios of those pairs of resistances are directly determined, the resistors of which have both been connected at two different times with the same source, that is to say, those pairs i, j $\in$ M of resistors, for which there are t $\in$ T, t' $\in$ T, and k $\in$ N such that con(i,k,t) and con(j,k,t') are true. Furthermore those ratios of resistances R(i)/R(j) indirectly obtain a numerical value from the "conditional definition", which can be expressed by a chain $\frac{R(i)}{R(j)} = \frac{R(i)}{R(i_1)} \cdots \frac{R(i_r)}{R(i_{r+1})} \cdots \frac{R(i_n)}{R(j)}$ of the directly determined ratios. All other ratios of resistances remain undefined and therefore we have to conclude that resistance is not definable in terms of electrical current in our theory, not even if R(1) = 1 is postulated by convention. Therefore resistance R has to be considered as a theoretical concept for our given antecedently available vocabulary.

The situation for voltage U is analogous. We may define it as follows:

$$t \in T \wedge t' \in T \wedge i \in M \wedge k \in N \wedge l \in N \wedge$$
$$\mathrm{con}(i,k,t) \wedge \mathrm{con}(i,l,t') \rightarrow \frac{U(k)}{U(l)} = \frac{I(i,t)}{I(i,t')} .$$

This expression is very similar to the corresponding one for resistance R, and in a very similar way we may conclude that voltage U is a theoretical term as well.

We now want to investigate the definability of resistance R and voltage U by a crypto-intensional function $\bar{I}$ instead of a purely extensional one. We obtain the crypto-intensional functions, if we replace the time argument in the purely extensional functions by a state argument. The states will simply be defined by ordered pairs $\langle i \in M, j \in N \rangle$, i.e. ordered pairs of resistors and sources. The state of the system is known, if it is known which resistors and sources are connected with each other. So our new typifications are

$$\bar{I} : N \times M \to \mathbb{R}; \quad \bar{R} : M \to \mathbb{R}; \quad \bar{U} : N \to \mathbb{R}.$$

The relation con is becoming trivial now since $con(i,k,\langle i,k \rangle)$, which is obtained by replacement of t by $\langle i,k \rangle$ in con, is analytically true. Therefore we can drop the condition in our conditional definition of $\bar{R}(i)/\bar{R}(j)$ and $\bar{U}(k)/\bar{U}(l)$ and obtain the result that these ratios are defined in any case. Thus if we stipulate $R(1) = 1$, and $U(1) = 1$, we can define resistance and voltage, and these concepts are no longer theoretical.

We have now seen in two examples that theoretical concepts become definable if we replace purely extensional by crypto-intensional functions. The list of such instances could be prolonged arbitrarily. There is however no point in continuing such an "inductive" process when one has understood what they have in common. We should come to a systematic discussion now, in which two questions will immediately arise:

1. What is distinctive about crypto-intensional functions, predicates, or relations in general, and what characterizes purely extensional concepts.

2. If we suspect that the possibility of defining theoretical concepts by crypto-intensional concepts is the reason for Ludwig's confidence in his claim, is there anything wrong with it? Does it become illegitimate by the application of crypto-intensional concepts?

The answer to the first question will help us to discuss the second. If we are looking for concepts suitable for the description of observations, even if we use scientific instruments for them, we should in any case chose our predicates, relations or functions, which describe physical situations or states of physical systems, in such a way that only names of physical objects, space, time or space-time points appear as arguments in them. These names refer to independently existing entities in the sense that their properties are logically independent. For time points it is the case that what happens at one time does not allow us to draw conclusions about any event at another time, as Hume already has noticed, as long as we do not make use of natural laws in our inferences. For physical objects it is also clear that one thing is not affected logically by the state of the other. In an ideally purely extensional language nothing which can be said about an ordered set of individuals makes a statement about another set of individuals meaningless. Let us imagine for example that the basic sets of our language are a time interval, an area in space, and a set of physical things. Thus we may say in this language that Mr. Meyers (m) occupation at $12^{00}$ p.m., Okt. $13^{th}$, 1980 ($t_o$), in his bed, Bismarckstr. 19 4500 Osnabrück, Germany ($x_o$), was sleeping (as). We express this by: as($m,t_o,x_o$). Let $x_1$ be "on the top of the Matterhorn". Then if as($m,t_o,x_o$) is true, as($m,t_o,x_1$) will be meaningless. Nothing can be said about Mr. Meyer's occupation at time $t_o$ on the Matterhorn, since he was not there at that time. We cannot, therefore, interpret a language, which contains the predicate as by two-valued truth functional semantics; it would be necessary to introduce a third value "indeterminate", as has been done for instance by U. Blau (1978). Such three valued semantics would establish an extensional interpretation of the language. If one does not like such an interpretation, one has to chose the arguments of predicates and functions in such a way that a true or a false elementary proposition with one set of arguments does not make elementary propositions with other arguments meaningless, and so one is led to purely extensional languages. There is, however, an interpretation which is no longer extensional. We could understand as($m,t_o,x_1$) in the following way: "If Mr. Meyer would have been put on the top of the Matterhorn at $12^{00}$ p.m., 10-13-1980 and nothing else would have been changed in the world, his occupation would have been sleeping." This is an interpretation by a counterfactual, which is made possible by the use of crypto-intensional predicates.

Now, why does'nt as(p,t,x) := "person p's occupation is sleeping at time t at the position x", assume truth values true or false for the real world for any combination of the variables p,t,x? These variables are not independent of each other. The variable x can be understood as a function of t and p. Then t and p are independent variables and x is a dependent variable, the value of a function. So the distinction between independent and dependent variables seems to be important. We define:

> Assume that in a language L describing some features of the real world there are variables of different types belonging to different basic sets $x_1 \in E_1,\ldots,x_n \in E_n$, $y_1 \in F_1,\ldots,y_m \in F_m$. Let there be functions in L for $i = 1,\ldots,m$: $g_i$: $E_1 \times\ldots\times E_n \rightarrow F_i$, such that the first order predicates in L obtain truth values "true" or "false" exactly for combinations of variables $x_1,\ldots,x_n,y_1,\ldots,y_m$ where $y_i = g_i(x_1,\ldots,x_n)$, $(i = 1,\ldots,m)$, then we shall say that $x_1,\ldots,x_n$ are independent variables in L and $y_1,\ldots,y_m$ are dependent variables in L.

There may be bijections in L which map the basic sets $E_1,\ldots,E_n$ and $F_1,\ldots,F_m$ on the real numbers or on ordered sets of real numbers. In such cases numbers or ordered sets of numbers can function as names for the individuals of the basic sets. The variables could then be considered as numerical or vectorial variables. We could replace such sets $E_i$ or $F_j$ by $\mathbb{R}$ or $\mathbb{R}^k$ , which are called auxiliary basic sets by Bourbaki. Nevertheless, one should not forget that physical functions or predicates do not talk about real numbers but about physical quantities.

We now can say that it is characteristic for crypto-intensional predicates that their arguments are at least in part dependent variables, i.e. values of the functions $g_1,\ldots,g_m$. This is of course also true for nomological predicates, as for instance for $ig(p,V,T) \leftrightarrow p \cdot V = R \cdot T$, which is true for ideal gases, where R is a universal physical constant, p the pressure of the gas, V the volume of 1 mol and T the temperature. Here p,V,T are altogether dependent variables. But these nomological predicates are completely given by natural laws and contain no information about peculiarities of individual physical systems or objects. We can therefore define:

> In a language L, which partly describes the real world, those predicates are purely extensional the arguments of which are only independent variables. Predicates which have one or some dependent variables as arguments, and which are not nomological are crypto-intensional.

Now there may be trivial cases, in which the functions $g_1,\ldots,g_m$ do not depend on the physical state of the world. These cases are not interesting for us here.

More important are cases where $g_1,\ldots,g_m$ are characteristic for certain possible worlds, and assume different values for the same arguments in different possible worlds. In such cases some atomic sentences may have truth values "true" and "false" in other worlds, even if they have no truth values in the real world. Let $x_1$ and $x_2$ be the positions of two charges in the real world. Then the proposition $f(x_1,x_2) = 10$ Newton, where f is the force acting from particle 2 on particle 1, has a truth value in the real world, while "$f(x_1',x_2') = 10$ Newton" for $x_1' \neq x_1$ and $x_2' \neq x_2$ has not. There may be a possible world, however, where "$f(x_1',x_2') = 10$ Newton" is true. And therefore we might read this formula in the following way: "If $x_1'$ and $x_2'$ would be the positions of the particles 1 and 2, the force f acting from 2 on 1 would be equal to 10 Newton".

Those nontrivial cases justify the expression "crypto-intensional", since in them a covert intensional interpretation by counterfactual conditionals becomes important in physics and can be used for defining otherwise theoretical terms.

We have to add a further definition of purely extensional and crypto-intensional languages:

> A language L is purely extensional, if it contains no crypto-intensional predicates, otherwise it is crypto-intensional.

So far the first question has been answered, and we have now to discuss the second: Is there something objectionable in Ludwig's application of crypto-intensional languages?

The examples presented above suggest that possibly all theoretical concepts are definable, if the antecedently available vocabulary is given in a suitable crypto-intensional form. If this is true, why should we not define theoretical concepts (in one of the above examples this was electrical charge) in such a language? Indeed we may do so if we like. We must ask ourselves however which thesis we have proved by such a procedure. Besides the crypto-intensional available vocabulary $L_{ci}$ (including in the example a state-dependent force function) a purely extensive available vocabulary $L_{pe}$ (containing in the above application a force function depending only on time and particle-number) describes the same data which are provided by methods of measurement made possible by the pretheories.

So there are two languages $L_{ci}$ and $L_{pe}$ which describe the observations. Some concepts of the language $L_T$ of the physical theory are <u>not</u> definable in $L_{pe}$ and are definable in $L_{ci}$. From this we conclude that some concepts in $L_{ci}$ are not definable in $L_{pe}$. (Thus the state dependend force function in the electrostatical example cannot be defined in its purely extensional description.)

And now we may draw the following conclusion: The definition of some concepts of $L_T$ is achieved by the use of a language $L_{ci}$ which contains concepts being themselves theoretical in $L_{pe}$. Such a procedure could be judged as a circumvention of the problem of theoretical terms, if we accept the following prescription: Describe the observational facts (the "Realtext" in Ludwig's terminology) in the poorest possible language! Since $L_{pe}$ is poorer than $L_{ci}$, according to this postulate the question, whether the concepts of $L_T$ can be defined, has to be posed for $L_{pe}$ and not for $L_{ci}$. But do we have to accept this prescription? Nobody can be forced to do that. Therefore at this stage of the discussion we are inclined to state the following result: The thesis that concepts of most physical theories cannot be defined in its antecedently available vocabulary cannot be upheld in its general and unspecified version. We have to restrict it in some way for example in the following manner:

The concepts of $L_T$ are not definable in $L_{pe}$ for most physical theories T.

This relativized thesis however loses much of the appeal which the general thesis still had, and enables Ludwig to say that what is usually considered as being a theoretical term can be defined in the observational language in which the "Realtext" is described. Thus Ludwig may successfully defend his thesis, which contradicts the "standard view" of a whole generation of philosophers of science, as long as inaccuary of measurement has not yet been taken into account, which shall be done in the next section.

## IV. FURTHER OBJECTIONS AGAINST THE DEFINABILITY OF THEORETICAL CONCEPTS ON THE BASIS OF THE INACCURACY OF MEASUREMENTS

The preceding discussion has come to the result that we may define what usually counts for theoretical concepts, if we apply crypto-intensional languages. And apparently there is no problem of theoretical terms for people who do not object to these languages. There are however other objections against theoretical terms of a different kind, which may be summarized in the following way: We do not conclude with certainty from observed data to values of theoretical functions; we are rather led to these values by inductive inferences. A definition would enable us in any case to derive logically any proposition about the definiendum from one about the definiens. Therefore we do not apply definitions of theoretical functions, when we determine their values by measurement.

Arguments of this type may refer to the theory of experimental errors. The measured values of physical quantities are not directly identified with their true values. We cannot obtain more from our data than a best estimate, which must not be mistaken for the real or true value. If repetition of a series of measurements yields a different value, we would not conclude that the true value of the quantity has changed. We would rather once more make use of the calculation of errors and obtain a still better estimate from all hitherto known data. Such a procedure is incompatible with the idea that physical quantities may be defined by their method of measurement.

A different version of this argument may be demonstrated by the example of intelligence measurement (R. Carnap 1956, section xi; W. Stegmüller 1970 IIB, p. 232 ff.). If a person p is given an intelligence test, the result is always open to objections of the following kind: p was under the influence of alcohol; p was very well acquainted with the test applied, and similar arguments. Therefore the test does not define the intelligence quotient. This second example is equivalent to the first in the following respect: In both cases the physical or psychological system is disturbed from outside, while the definition of the quantity presupposes an undisturbed system. One may ask, if the possible intervening variables might be accounted for in an improved operational definition of the quantity. Thus we may begin our definition of the intelligence quotient by the phrase: "if the person p is not under the influence of alcohol or any other drug, if p is not tired or stressed, etc. ..." and by this procedure we may indeed obtain a better definition. This method of improvement however does not work for the statistical perturbations leading to statistical errors. For these influences are both unknown and uncontrollable. It would be pointless to define a quantity by measurements which are free from statistically irregular influences, since there are no such measurements as we are taught by laboratory experience. We are therefore forced to consider any measurement as an inductive and not as a deductive inference, which cannot be transformed into a definition of the measured quantity.

One may ask how this argument may look in Ludwig's system of theory reconstruction. If Ludwig has been aware of experimental errors, they must appear in his reconstruction scheme in some way. And indeed they do. Ludwig accounts for them by his imprecision sets (1978, p. 50). Thus we are now led to one of the central themes of this volume, and therefore it may be unnecessary to give here an introduction to the concept of inprecision set. The reader may find the needed information for example in the contribution of C.U. Moulines (see definition D1), and also in G. Ludwig's own paper to this volume.

Now, presupposing some familiarity of the reader with imprecision sets, we ask whether Ludwig's application of imprecision-sets to the logical reconstruction of physical theories is compatible with the definability of all terms of a theory in the antecedently available vocabulary. Thus we discuss the preceding argument of undefinability due to imprecision of measurement in Ludwig's terms.

We shall introduce imprecision sets directly on the set $M_{pot}$ of possible physical processes which are described by the system. Let us assume that we are dealing with a theory in which such processes are described as functions of time into the state space S. Such a function describes the time evolution of the physical state. The state space S is the set of all possible momentary states of a physical system. Therefore, when T is a time interval, physical processes are described by functions $s : T \rightarrow S$; so we have

$$M_{pot} = \{s \mid s: T \rightarrow S\}.$$

J.D. Sneed calls the set of "possible models" of a physical theory $M_o$ or $M_p$. A possible model is very similar to a physical process $s \in M_{pot}$. Let $s \in M_{pot}$ be such a process and let $E_1, \ldots, E_n$ be the basic sets or "domains", as Sneed calles them, of the theory, then the ordered set $\langle E_1, \ldots, E_n, s\rangle \in M_p$ is a possible model. (See also C.U. Moulines summary in the introduction of his paper in this volume.) In Ludwig's terminology $M_{pot}$ is the set by which the structure term s of the theory is typified: $s \in M_{pot}$.

To any physical process $s \in M_{pot}$ there belong a "neighbourhood" will belong V(s) of processes which cannot be distinguished from it by measurement. We may introduce V as a function from $M_{pot}$ into $(M_{pot})$:

$$V : M_{pot} \rightarrow \text{pot}(M_{pot})$$

For $t \in V(s)$ we can also write $\langle t,s\rangle \in U$, V(s) being $V(s) = \{z \mid \langle z,s\rangle \in U\}$, and U having the typification

$U \in M_{pot} \times M_{pot}$. U is an imprecision set on $M_{pot}$, and we demand that U belongs to a set N ($U \in N$), which satisfies the axioms for a uniform structure (Ludwig 1978, p. 54,55, C.U. Moulines in this volume, definition D1). Moulines introduces the imprecision-sets as subsets of $M_p \times M_p$, while in my presentation they are subsets of $M_{pot} \times M_{pot}$. This difference is unimportant, however, since there is a one-to-one correspondence between the elements of $M_p$ and $M_{pot}$. The imprecision set introduced here differs from Ludwig's inaccuracy sets, since it is defined as a subset of $M_{pot} \times M_{pot}$, while Ludwig defines them for cartesian squares of basic sets $E_1,..,E_n$, i.e. sets of individuals the theory is talking about. But one can show that imprecision sets over the basic sets generate corresponding imprecision sets over all echelon terms of them, i.e. over all sets which might be obtained from them by successive application of the operations $\times$ and pot. Thus we are led in any case to the definition of an imprecision set U over $M_{pot}$, $U \subset M_{pot} \times M_{pot}$.

Next we have to write down the mathematical part of the physical theory. Let $E_1,...,E_n$ be the basic sets of the theory, the sets of physical objects, space, time, or space-time points which are appearing in the theory. We may then write down the main axiom of the theory as follows

$$P(E_1,...,E_n,s)$$

This axiom claims too much about physical processes. Since our knowledge about them is imprecise, we can only say the following: If $s \in M_{pot}$ is a physical process, there is a process $s' \in M_{pot}$ in its neighbourhood V(s) which satisfies the equations of the theory. We obtain a logical expression which might be abbreviated by $\tilde{P}(E_1,...,E_n,s)$:

$$\tilde{P}(E_1,...,E_n,s) \iff \vee s'(s' \in V(s) \wedge P(E_1,...,E_n,s'))$$
$$\iff \vee s'(\langle s',s\rangle \in U \wedge P(E_1,...,E_n,s'))$$

The expression $\tilde{P}(E_1,...,E_n,s)$ might be interpreted as axiom of the "smeared out" theory, the theory which really can be known, while the axiom of the exact theory $P(E_1,...,E_n,s)$ contains an excess of "information" which has no empirical significance.

It is hard not to notice immediately that $\tilde{P}(E_1,...,E_n s)$ has the form of a Ramsey sentence, in which s' is the Ramsey variable*).

*) See Stegmüller 1970, Vol. IIC, p. 400-437. Ramsey conjoined the axioms of a physical theory and replaced its theoretical predicates by predicate variables (Ramsey variables), which are bound by existential quantifiers put before the whole expression. The "Ramsey sentence" thus obtained, contains only observational predicates as logical constants. Therefore one can say that it is an expression for the theory in the observational language.

This means, if we replace the variable s' by a constant $s_o$, $s_o$ will be a theoretical concept. Since $s_o$ contains all functions which describe the physical process in question, these turn out to be theoretical terms if we cannot find a definition for them in terms of the functions used for the formulation of $s_o$. If for example $s_o$ is the time evolution of the electromagnetic field, s will be the measured time evolution of the field, while $s_o$ characterizes the true evolution of it. Since the relation between s and $s_o$ is given by $s_o \in V(s)$ or $\langle s_o, s\rangle \in U$, there seems to be no way to define $s_o$ in terms of s. Any $s' \in V(s)$ may be the true $s_o$, the true description of the physical process.

We come to the conclusion that the well known argument for the existence of theoretical terms finds a nice representation in Ludwig's reconstruction, and we may see in it very clearly that physical quantities cannot be defined by their measurement procedures, if we apply Ludwig's imprecision-sets as a tool for our argument. Thus our discussion shows at least that for Ludwig there must also be theoretical concepts in a certain sense.

Did Ludwig not notice that the true physical quantities cannot be defined by their measurement values? He did, indeed under the heading "idealization":

> "Die Anwendung der mengentheoretischen Axiome ... entspricht also einem für die theoretische Physik grundlegend wichtigem Vorgehen, dem wir schon in bezug auf die Sachlage der unscharfen Abbildungen begegnet sind: der Idealisierung. Das soll heißen: Ohne daß durch die benutzten Realtexte eine physikalische Frage entschieden werden kann, werden Axiome der Art aufgenommen, daß diese weder den Realtextstücken zu widersprechen scheinen, noch von ihnen kontrolliert werden können. Diese Axiome malen sozusagen $MT$, das mathematische Bild, feiner aus an Stellen, wo man eigentlich vom Realtext her nicht weiß, 'wie es dort weitergeht'."(1978, p. 106.)

This is quite in agreement with what has just been said about an excess of empirically insignificant information given by physical theories. Therefore I see no reason why Ludwig should not finally accept the thesis that theoretical concepts in physics are indispensable.

## LITERATURE

U. Blau 1978, Die dreiwertige Logik der Sprache, Berlin/New York: de Gruyter

N. Bourbaki 1968, Theory of Sets, Paris: Hermann, and Reading (Mass.): Addison Wesley.

R. Carnap 1956, "The Methodological Character of Theoretical Concepts", Minnesota Studies in the Philosophy of Science Vol. I, Minneapolis (Minn.): Univ. of Minnesota Pr.; p. 38-76.

M. Heidelberger 1978, "Über eine Methode der Bestimmung theoretischer Terme", in W. Balzer, A. Kamlah (eds.) : Aspekte der physi-

kalischen Begriffsbildung, Braunschweig/Wiesbaden: Vieweg; p. 37-48.
C.G. Hempel 1973, "The Meaning of Theoretical Terms: A Critique of the Standard Empiricist Construal", in Logic, Methodology and Philosophy of Science IV (ed. by P. Suppes et.al.) Amsterdam/London/New York: North Holland Publ. Comp./American Elsevier Publ.Comp.
A. Kamlah 1977, "Erläuterungen" for Philosophie der Raum-Zeit-Lehre, Vol. 2 of Hans Reichenbach, Gesammelte Werke, Braunschweig/Wiesbaden: Vieweg; p. 389-432.
G. Ludwig 1974, Einführung in die Grundlagen der theoretischen Physik, Vol. 1, Raum, Zeit, Mechanik, Düsseldorf: Bertelsmann.
G. Ludwig 1978, Die Grundstrukturen einer pyhsikalischen Theorie, Berlin/Heidelberg/New York: Springer.
G. Ludwig 1978a, "Axiomatische Basis und physikalische Begriffe", in W. Balzer, A. Kamlah (Hrsg.): Aspekte der physikalischen Begriffsbildung, Braunschweig/Wiesbaden: Vieweg; p. 99-107.
G. Ludwig 1981, "Axiomatische Basis einer physikalischen Theorie und theoretische Begriffe", in: Zeitschrift f. allg. Wiss.theorie
H.W. Schürmann 1977, Theoriebildung und Modellbildung, Wiesbaden: Akad. Verlagsges..
J.D. Sneed 1971, The Logical Structure of Mathematical Physics, Dordrecht: Reidel.
W. Stegmüller 1970, Probleme und Resultate der Wissenschaftstheorie und analytischen Philosophie, Studienausgabe, Vols. IIB and IIC, Berlin/Heidelberg/New York: Springer.

# ABSTRACTION, IDEALIZATION AND APPROXIMATION

## A Reflection on the Nature of Scientific Concepts

U. Majer

Philosophisches Seminar der Universität
Nicolausberger Weg 9c
3400 Göttingen
Federal Republic of Germany

It has recently been claimed by M. Dummett[1)] and other that to be a realist implies the acceptance of Aristotle's principle of the excluded middle that every proposition has one of two truth-values, either true or false. Hence, if there are propositions the truth-values of which are not decidable in principle, anti-realism perhaps in the form of intuitionism or verificationism is epistemologically a more reasonable position than unrestricted realism presupposing verification-transcendent truth-conditions.

It is not my aim to discuss the argument in detail; all that I want to do is to compare two different conceptions of concepts against this background, thereby throwing some light on the nature and role of concepts in science. Specifically, I have two proposals in mind, namely that of G. Frege [2)], taking concepts as functions in judgements, and that of the 'picture-theorists' - beginning with Hertz[3)] and Mach[4)] and culminating in the mathematically sophisticated work of G. Ludwig[5)] - taking concepts as instruments for the description of the physical world, which means as more or less accurate representations of the structure of the physical phenomena (i.e. measurements).

## I. FREGE'S CONCEPTION OF CONCEPTS[6)]

In the spirit of the above definition - and not only in that - Frege was undoubtedly a realist because he insisted on the unrestricted exactness of all scietific concepts, not only the logical but also the descriptive (empirical) ones. 'Exactness" here means, mathematically speaking, that a concept is a characteristic function - indeed a total one - from a domain D of objects into a two-valued range of constants, the True and the False, such that for every object, pair of objects, and generally n-tupel of objects,the value of the function is either the True or the False, according to whether, the object in question, respectively pair of objects etc. falls under the concept or not. It is crucial for an understanding of Frege's conception of concepts that every concept-expression expresses a sense (Sinn) which is epistemologically prior to the characteristic function the expression designates; that is to say that the sense of the concept-expression as a constituent of scientific thoughts determines in a unique way the characteristic function which is the 'Bedeutung' of that expression. That means that we know the function and thereby its course of values only by reflection - Frege says by judgement - on the sense of all saturated expressions, i.e. of sentences we obtain, by substituting different names of objects in the argument-place of the concept-expression. Respectively, the extension or what Frege calls the 'Umfang' of a concept is uniquely correlated with the course of values of the characteristic function in such a way that those and only those objects, pairs of objects etc. "fall under the concept" (are elements of the extension of the concept) for which the value of the function in judging a thought is 'the True'; hence the extension of a concept is the set of all n-tupels of objects for which the value-course of the function is 'the True'. Examples are not hard to come by: Logical concepts (of first order) are such characteristic functions under which only truth-values, pairs of truth-values and in general n-tupels of truth-values fall; hence the extension of which consists of truth-values and truth-values only. On one of these functions, namely the 'identical'-representation of the concept of 'Bejahung', which is expressed by the horizontal stroke "——", Frege builds up his logical system of the "Begriffsschrift": Under this concept falls 'the True' and only 'the True' because its value is 'the True' iff its argument is 'the True'. Therefore Frege calls this concept also the concept of truth. Descriptive concepts (of first level) are such characteristic functions under which only objects (of usual kind), pairs of objects and such like fall, excluding truth-values; hence the extension of which consist always of n-tupels of objects of one or another kind, except truth-values. For example, under the concept 'euclidean triangle' fall all geometrical objects, the sum of whose three angles is equal to $180^{o}$ degree. (If this explanation of the concept 'triangle' or its extension respectively seems vacuous or analytical, read the following passage carefully.)

There is a great lacuna in Frege's writings in as much - so far as I know - Frege never stated explicitly what the sense of a concept-expression is, nor what it should be. He only insisted that they have a sense, at least in the context of sentences so far as these are scientific. Other than in the case of names, he simply took it for granted that concept-expressions have sense - indeed an objective or absolute one - which we can grasp and make the basic of our judgement. He concentrated solely on the question what the 'Bedeutung' of concept-expressions could and should be; (whereas in the case of names the question of sense and not of reference (Bedeutung) caused the main trouble (see next passage)). Now, it is plain enough that if the 'Bedeutung' of a concept-expression is the total characteristic-function called concept, then its sense can nothing be else than the "way in which this function is given", that is the way in which its course of values for a certain range of objects is determined. But that in turn is nothing else than the 'Merkmale' of the concept in question by which we predicate corresponding properties, relations etc., or 'Eigenschaften' in general to certain objects, pairs of objects etc., if we use the concept-expression to make assertions: asserting thoughts of which we have already judged that they are true, that the value of their combined characteristic-functions is 'the True'. This interpretation is confirmed not only by a careful reading of Frege's writings, especially the 'Grundlagen' and 'B. u. G.', but also by means of the philosophical tradition out of which Frege developed his logico-philosophical position, namely that of Kant, Trendelenburg[7] and Lotze[8]. For these the 'contents' - or 'sense' in Frege's later terminology - of a concept-word was the sum of the 'Merkmale' as they were combinated in the concept. What Frege added to the tradition in this respect was only the sharp distinction between the 'Merkmale' of a concept and the 'Eigenschaften' of objects: 'Merkmale' are always the characteristic features of concepts, whereas 'Eigenschaften' are the properties, relations etc. of objects only. Therefore, the latter can be predicated by the former to objects, but never to concepts, at least not in the same logical way. If the predication is true, that means if the object has the properties specified by the 'Merkmale' of the concept, the object in question falls under the concept. But concepts (of the same level) are at best subordinated to another, namely 'a is subordinated to b', if the set of the characteristic features of a contains that of b. If we none the less want to assert something about concepts of first order we have to introduce second order concepts, because the difference between objects and concepts is irreducible and what can be said of objects never can be said of concepts, and vice versa. So for example, the assertion that under a certain concept, let's say that of a 'round-square', no object falls can only be expressed by the second order concepts of 'quantification', generality or existence as you like, because it is not a proposition about objects, at least not immediately, but about the concept 'round-square', namely that its characteristic features 'round' and 'square' as combined in that concept contradict each other by the very meaning (Sinn) of the expression. Hence, there cannot 'exist' a single object which has both propert.es: 'round' and 'square'. It is important to note that this judgement about the concept 'round-square' is not an empirical one because it doesn't depend on the (cardinality of the) range of objects, which the concept can take as arguments, rather the conclusion follows immediately from a reflection on the sense of the concept-expression. So it looks as if the judgement were an analytical one; but we have to be careful in this respect too, because the existence of such an obscure object as 'round-square' is not excluded logically, but only in virtue of our geometrical 'intuition'. Why then is 'generality' (or 'existence') a logical concept (of second order) at all? The reason is simply this: the value of the second-order function 'generality' depends only on the truth-values of the elementary thoughts which can be formed out of the first-order concept 'round-square' (as well as any other concept).

Now we understand better what the role of concepts in science is: according to this conception of concepts, they are functions in judgements, characteristic functions which uniquely determine the extension of concepts. But if we ask judgement of what, the somehow meagre but none the less correct answer is: judgements of thoughts. If we ask further what kind of entities thoughts are, the only answer we get from Frege is that they form a third world of entities which are either true or false once and forever, independent of our thinking, i.e. of our grasping and judging them. At this somewhat surprising point, I think it is helpful to remember that thoughts form a part of the sphere of sense and that 'sense', at least in its ethymological meaning, is an epistemic category which is related with our 'sensual' perception' or perhaps more revealing in German: 'sinnliche Wahrnehmung'. Now two circumstances are crucial for an understanding of Frege's epistemology, and hence his 'meaning-theory' as Dummett would say: First of all, that we only know the 'Bedeutung' of an expression <u>via</u> its sense; the sense is 'the way of giveness' of the 'Bedeutung' not only of objects but also of concepts; there are neither objects nor concepts 'per se', but only judgements can reveal what the 'Bedeutung', the real <u>essence</u> behind the surface of sense is. Secondly, and of no less importance is the circumstance that there is no canonical 'way of giveness', neither of objects nor of concepts; every object as well as concept can appear in a multiplicity of senses; all that we can do is to combine them tentatively in thoughts - the way in which objects are given with the 'Merkmale' of concepts - and ask which of these tentative thoughts are true; only by judgements with respect to the truth, the eternal truth of thoughts, we can recognize what the common 'Bedeutung' behind the multiplicity of senses is.

The answer to the last question shows in what sense Frege constructs his 'semantic theory': Semantic in his sense has nothing to do with the 'interpretation' of languages as it is understood nowadays, for example in 'model-theory', taking languages as uninterpreted formal systems which receive their 'interpretations' by presupposing a relational structure - given from heaven - and specifying 'projection-rules'. Rather it is an 'interpretation' in a very peculiar, indeed metaphysical sense: Granting that we can grasp thoughts, he asks what is the ontological 'Bedeutung' of their constituents, if certain thoughts will be true definitively. The first quite general answer is: the 'Bedeutung" are certain kinds of objects and concepts, the first are self-contained, the second are in need of completion (by the first).

## II. THEORIES AS CONCEPTS

In the following I take physical theories as concepts in the Fregean sense, that is, as very complex concepts which are composed out of other concepts of first and second order under which certain sets of physical objects, sets of pairs of physical objects and in general sets of n-tupels of physical objects fall. This has two obvious consequences: First, a physical theory is not an uninterpreted formal, or a pure mathematical i.e. set-theoretical language which acquires a physical meaning by interpretation-rules. It is not a language at all; rather it is a complex characteristic function with a definite sense, namely a certain combination of 'Merkmale' we use to predicate corresponding properties to physical objects. Secondly, a physical theory is not and is not expressed by a set of (closed) sentences which are either true or false; rather it is an 'unsaturated' concept in need of completion by arguments. Its sense, a certain combination of 'Merkmale' can be used in application to certain tupels of objects, resp. the ways in which they are given to form true or false thoughts, expressed by sentences. This does not preclude that the formulation of the theory contains certain forms of open sentences, so called equations between physical concepts (functions) which are the physical laws of the theory. Nor does it preclude that the formulation of the theory contains certain kinds of names which designate the so called 'universal-constants' that are characteristic for the theory, like Kepler's 'constant' or Newton's 'gravitational-constant'. It only precludes that the theory itself and its laws refer to individual objects and constants which are contingent in respect to the general law - like applications of the theory. To give an example, not too trivial: Under the concept of Kepler's theory - let's call it a 'Keplerian-system' - fall certain sets of objects (planets) and others not, namely exactly those whose movements satisfy the three laws of Kepler. With this concept in mind we can form an indefinite set of thoughts, an element of which is expressed by the sentence: "The system 'sun-earth-moon' is a Keplerian-system", and this thought is obviously false because the moon does not move on an ellipse around the sun, as the sense the first law demands. Hence, the system 's-e-m' as a whole does not have the property of being a 'Keplerian-system', or what comes to the same does not fall under the concept of Kepler's theory. Of course, there can be other more complicated reasons why a system of planets does not fall under the concept of Kepler's theory, for example a violation of the second law; but in general it is sufficient that one of the 'Merkmale' of the concept is not fulfilled by the properties of the system.[9)]

## III. OBJECTIONS TO FREGE'S CONCEPTION

I have explained Frege's conception of 'exact' concepts at some length because it is the target of many objections, most of them taken from ordinary-language arguments.[10)] Insofar as these deal with context-dependet ambiguities of ordinary expressions like 'bank' or indexicals like 'he', 'she', 'it' etc. or tense-words like 'yesterday', 'today', 'tomorrow', and so on, I doubt their relevance. It is absolutely clear from Frege's writings that he never intended his conception of concepts as functions in judgement as a description of every-day-thinking, much less of our ordinary language. Rather it was intended as a normative rule of how we must think if we shall obtain the goal of science, that is the knowledge of trues.

Hence, the error of these objections lies in neglecting the fact that we have to disambiguate such context-dependent expressions according to the rule that the resulting thought is absolutely true or false, that means true or false without referring to certain space-time-intervals or other contingent circumstances. This we have to do before constructing a formal language (of any kind) and evaluating the truth-values of its sentences; I think in science we really do this.

Above and beyond all these pointless objections there is one argument to be taken very seriously coming from the epistemology of physics, which runs roughly such: Either physical theories taken as concepts are exact (characteristic functions), then it seem in the light of all experience that no real object falls under them, hence no sentence formulated with these concepts is true (the concepts have no real application at all); or the nature of physical concepts must be different in principle to ensure that they have at least some applications to 'reality'; hence, they cannot be characteristic functions from a domain of objects into the range of two truth-values.

The answer given in response to this dilemma are quite different depending on which horn of the dilemma one chooses. Either one modifies the notion of truth for empirical theories into 'probably true' or 'approximately true' or something like that, leaving physical concepts 'exact' in the sense of the acceptance of Aristotle's principle of bivalence for elementary propositions, deducible from the theory. Or else on changes the conception of concepts making concepts 'inexact' in the sense of a rejection of Aristotle's principle for elementary propositions, leaving the notion of truth for logical concepts untouched (except for a necessary but harmless 'technical revision'. Or else one alters both, the conception of truth and the conception of concepts by introducing respectively a new logic and foundation of semantic.

For instance, Popper and others have introduced the notion of 'verisimilitude' for theories leaving concepts 'exact' and modifying the notion of truth such that theories are more or less good approximations to the truth - the total truth possible - depending on the difference between true and false statements deducible from the theory. Similarly, Carnap has invented the notion of 'inductive confirmation' for theories (hypothesis) again leaving concepts 'exact' and defining a measure-function 'c (h,e)' for the confirmation of the hypothesis (h) through experience (e), which is in essence a conditional probability. Again, 'quantum-logicians' as well as intuitionists have given up, for rather different reasons, Aristotle's principle of bivalence and introduced a 'new' logic and foundation of semantic.

All these proposals I don't want to discuss, instead the answer I am interested in is the solution making physical concepts (theories) 'inexact' by introducing 'imprecision-sets' for concepts resp. their extensions. More specifically, the solution on which I shall concentrate is the approach of 'uniform-structures' as worked out by G. Ludwig and the Marburg-school of physicists. However, there is an expositional difficulty: Because I am not quite sure what the philosophy is behind this approach, I should first give an outline of the 'technique' and then discuss the merits of that approach. Since, the 'technique' is not quite intelligible 'per se', I will proceed in the following revised order: First, I'll give some information about the 'picture-conception' of theories as developed as physicists in the late 19th century, which is the historical origin and epistemological background of Ludwig's approach. Then I shall present the 'technique' in greater detail such that it is easy to translate it into a more convenient form of modern semantic. Subsequently, I'll ask how far this approach is a solution to the supposed problem of the dilemma. Insofar as the answer is negative at least in some respect, mathematical rigour and physical insight notwithstanding I'll suggest a way out of the dilemma from the judgement-view of concepts, reflecting on the role of 'abstraction', 'idealization' and 'approximation' in scientific concept-formation.

## IV. THE 'PICTURE-CONCEPTION' OF THEORIES

When in the second half of the 19th century the theory of electro-dynamics appeared upon the horizon, physicists like Kirchhoff, Hertz and Mach renewed distrust in the so called 'forces at a distance' which like gravitation, acted at a distance instantaneously. Of course, the argument was not so much that such forces could not 'exist' in the sense of providing a good description of the movements of the planets, rather the point was that these forces were themselves in need of a mechanical explanation and that no such explanation could be given in mechanical terms of push and pull alone. Hence, these forces could not serve as a causal explanation of the movement of bodies.[11] Newton's 'hypothesis non fingo' again became a strong argument against causal explanations of any kind whatever. Boltzmann[12] justifiably critisized Kirchhoff's efforts to eliminate the concept of forces from mechanics: "In half a page, forces had been defined away and banished from nature and physics made into a descriptive science properly speaking". Descriptions instead of explanation became the slogan of those days - up to the present time! The point was serious because with it the whole epistemology of science, especially of physics, changed radically: Instead of seeking an understanding of the essence of nature beyond the level of phenomena or 'sense data' by giving an explanation of the experimental results according to certain concepts at a deeper level, i.e. categories like substance and causality; quite the contrary became the goal of science, namely the restriction to the level of phenomena, avoiding all 'metophysical' concepts in science. Predictive or better deductive power in respect to novel phenomena became the single criterion for the correctness of logically consistent theories. Descriptions in mathematical terms and equations of all past as well as future experimental results had become the sole aim and purpose of theories; there was no room and no need for deeper going explanations whether physical or 'meta-physical'.

Out of this anti-explanatory, i.e. anti-metaphysical climax grew the 'picture-conception' of theories which is best stated in the well known remark of H. Hertz in his introduction to the Principles of mechanics: "We form for ourselves images or symbols of external objects, and the form which we give them is such that the necessary consequences of the images in thought are always the images of the necessary consequences in nature of the things pictured". No long commentary is required to understand the intention of Hertz's remark: There is a fundamental relation or correspondence between the external objects plus their relations, in short the relational structure of the external world, and our representations of the relational structure in mind by images or symbols such that the consequences of both stand in a one-one correspondence;namely the physical necessary consequence of the relational structure of the physical world and the logically necessary consequences of our images or symbols in thought. If this requirement is fulfilled the images or their expressions, the theories, are correct, and nothing beyond that structural isomorphism can be known, i.e. not what the 'true' nature or essence of the things pictured is.

From this picture-conception of theories it is only one step to Ludwig's general view of physical theories: Instead of 'mental' pictures or symbols in 'thought', which have in turn to be expressed by a language, spoken or written, we choose immediately for the representation of the relational structure of the physical world 'mathematical' symbols or terms which are part of the alphabet of a formal language we use to express the theory. According to Ludwig's terminology a physical theory 'PT' is formed out of three parts: a mathematical theory 'MT', a domain of reality (Wirklichkeitsbereich) 'W' and a set of correspondences '(——)' between 'MT' and 'W', such that, roughly speaking, a PT is equal to MT (——) W. The first step in constructing the correspondences between MT and W consists in selecting a proper subpart G of W, called 'basic domain' (Bereich realer Gegebenheiten) - or more specifically a concrete physical situation (experiment) of G - and co-ordinating certain symbols $a_1 \dots a_n$ of a formal language L with the elements of G such that there is a one-one correspondence between the elements of G and the 'names' $a_1 \dots a_n$; this co-ordination is conventional. Not conventional in contrast is the next step of stipulating "correspondence-rules"(c.r.), which accordin to Ludwig regulate the formation of the 'axioms of representation' (Abbildungsaxiome)[13] in knowledge of MT and a co-ordinated 'Realtext' (collection of facts) which denotes a real physical situation of G. Three general features of the 'correspondence-rules' can be formulated in advance (of any physical situation):

1) First, the c.r. select certain terms $Q_1,\dots,Q_p$ of MT, which are called 'picture-terms' (Bildterme)[13]
2) Second, the c.r. select certain relations $R_1(x_{\alpha_1},x_{\beta_1},\dots,\gamma_1),\dots$ $\dots, R_s(x_{\alpha_s},x_{\beta_s},\dots,\gamma_s)$ of MT, called (Bild-relationen)[13], which are in general n-place relations with n-'free-variables', for which the elements of $a_1,\dots,a_n$ can be substituted as arguments of the relations. Some of these relations may contain free-variables $\gamma_i$ for real numbers, $\alpha_{\gamma_i} \in \mathbb{R}$.
3) Third - and this is the first physically decisive step - the c.r. specify rules by which the symbols $a_1,\dots,a_n$ (co-ordinated with a 'Realtext') have to be typified according to axioms of the form:

(1) $(\text{——})_{r_1} : a_1 \in Q_1,\dots,a_n \in Q_n$

(2) $(\text{——})_{r_2} : R_1(a_{i_1},\dots,a_{k_1},\alpha_{\gamma_1}),R_2(a_{i_2},\dots,a_{k_2},\alpha_{\gamma_2}),\dots$

$\neg R_1(a_{j_1},\dots,a_{l_1},\alpha_{\gamma_1}),\neg R_2(a_{j_2},\dots,a_{l_2},\alpha_{\gamma_2}),\dots$

Intuitively speaking, what the c.r. regulate is the setting of a 'categorial frame' in which the symbols $a_1,\dots,a_n$, co-ordinated with a 'Realtext', are correlated with mathematical terms and relations in such a way that the last ones represent 'real physical situations' consisting of certain types of 'objects' standing in certain types of relations, and not in others.

The result of all this mathematical formalism is, as Wittgenstein[10] says in the 'Tractatus': "Durch den ganzen logischen Apparat hindurch sprechen die physikalischen Gesetze doch von den Gegenständen der Welt' (6.3431). Yet it is important to understand that the c.r. only characterize the form of the typifying axioms and not the axioms themselves, which can be formulated only in the knowledge of at least one concrete physical experiment. Hence, the correspondence rules only specify rules for possible frames of categories which can be used in the light of empirical knowledge to formulate a set of 'basic' propositions, the axioms of the observational report. (These propositions are basic insofar as their 'logic' only contains "affirmation", "negation" and "conjunction", but no quantifiers and no modal-operators.) Till now, not much has happened.

But, if once the axioms are formulated, one of two things can happen: Either the axioms are consistent, (i.e. in connection with MT they do not lead to a contradiction) then the resulting PT is a "useful" theory; "describes the facts useful" or the axioms are inconsistent, (i.e. lead in connection with MT to a contradiction) in which case the PT is "not-useful".

Now, it is one of Ludwig's basic convictions that nearly every PT leads quickly to contradictions by extending the observational report; not so much that its axioms alone contradict each other, rather the reason is that we have not choosen the 'right' MT for the description of facts. It is an 'inadequate picture of reality'. This is in Ludwig's terminology the core of the dilemma in applying 'exact' concepts to reality. Now, what is the cure for the dilemma? For this on must remember that Ludwig rejects 'ex-cathedra' every solution changing the 'logic' of basic propositions, either the so-called 'probability'-solution, making basic propositions more or less 'true' according to a certain probability, or the quantum-logic-approach, in which the truth-value of basic propositions depends on other propositions. Like Popper and Carnap and unlike the quantum logicians he insists that at least basic propositions have to be definitively true or false. Being, in principle sympathetic with that position I ask what is the solution to the problem.

## V. THE TECHNIQUE OF 'IMPRECISION-SETS'

The basic idea of the solution is quite simple and in principle well known from modern semantics, although Ludwig's approach is rather complicated in detail. Instead of giving every relation-sign just one extension, it is interpreted in different ways, such that the resulting 'extension' is a set of extensions, connected by a 'family-relation' $\mathcal{R}$, where $\mathcal{R}$ usually is reflexive, symmetric and transitive.

That means in Ludwig's approach that we have to replace the 'exact' or 'ideal' picture-relation[13] $R_\mu(x,y,\ldots,\alpha)$ in the axioms of representation[13)] by a set of 'imprecise relations' $\tilde{R}_{\mu}^{U}(x,y,\ldots,\alpha)$, which are defined by the 'ideal' relation $R_{\mu}(x,y,\ldots,\alpha)$ and a 'imprecision-set' U, such that for every tupel of element $\langle x,y,\ldots,\alpha\rangle$ occuring in the ideal relation $R_{\mu}(x,y,\ldots,\alpha)$ there exists another tupel of elements $\langle x',y',\ldots,\alpha'\rangle$ standing in the same relation, and the pair of both tupels is an element of U:

$$(1)\quad (\exists x')(\exists y')\ldots(\exists\alpha')[R_{\mu}(x',y',\ldots,\alpha') \wedge (\langle x',y',\ldots,\alpha'\rangle,\langle x,y,\ldots,\alpha\rangle) \in U]$$

Hence, the decisive step for the selection of a set of imprecise relations $\tilde{R}_{\mu}^{U}(x,y,\ldots,\alpha)$ is the construction of an adequate imprecision-set U for every relation, or immediately more advanced, the construction of a system $N_{\mu}$ of 'more and more refinable' imprecision-sets $U \in N_{\mu}$, such that according to Ludwig the system $N_{\mu}$ fulfills the following five conditions of a so-called 'uniform-structure'. [Of course, because the $U \in N_{\mu}$ are subsets of $M_{\mu} \times M_{\mu}$, where $M_{\mu}$ is defined as the set of all possible tupels of elements, that can occur in $R_{\mu}(x,y,\ldots,\alpha)$, the product-set $Q_1 \times Q_2 \times\ldots\times \mathbb{R}$, $N_{\mu}$ is a subset of the power-set of $M_{\mu} \times M_{\mu}$, $N_{\mu} \subset \mathcal{P}(M_{\mu} \times M_{\mu})$]. The first three of the five conditions on $N_{\mu}$ are the previously mentioned conditions of reflexivity, transitivity and symmetry for the elements U of $N_{\mu}$. (Compare Ludwig (1978), page 53/54). The last two conditions are the decisive 'idealizations' for the formation of uniform-structures:

4) refinement-condition: If $U_1 \in N_{\mu}$ and $U_2 \in N_{\mu}$, then also $U_1 \cap U_2 \in N_{\mu}$, which roughly means that the intersection of two 'useful' imprecision-sets leads as a 'refined' imprecision-set, and to a 'useful' description of facts.

5) condition of unrestricted refinement: For every $U \in N_{\mu}$ there exists a $V \in N_{\mu}$ such that $V^2 \subset U$, where $V^2$ is defined as: $V^2 \equiv \{(z_1,z_2)\mid$ such that there is a z with $(z_1,z)\in V \wedge (z,z_2)\in V\}$, which means intuitively - I quote - "U contains V twice".

If we now add to every 'ideal' picture-relation[13)] a corresponding uniform-structure $N_{\mu}$, we can define for each $R_{\mu}(x,y,\ldots,\alpha)$, according to (1) a 'uniform-set' of imprecise-relations $\tilde{R}_{\mu}^{U}(x,y,\ldots,\alpha)$ which in the 'limes of refinement' exactly contains the 'ideal' picture-relation[13)] $R_{\mu}(\quad)$. If we further replace the 'ideal' relations $R_{\mu}(\quad)$ in the c.r. by the thus defined uniform-sets of imprecise-relations $\tilde{R}_{\mu}^{U}(\quad)$ we obtain the 'imprecise-correspondence rules', furnished with 'possible pictures of reality'. The point of this technique of working with uniform-structures is, that we now are in a position to formulate the 'axioms of fuzzy-representation'[13)] such that:

1) the basic propositions are 'practically' true. That means they are true in a certain range of 'approximation' determined by the 'finest' imprecision-set U possible, and

2) the basic propositions together with MT lead to no contradictions, or perhaps better, every contradiction can be avoided by selecting more 'coarse-grained' sets of imprecise-relations as pictures of reality.

Coming back to the remark at the beginning of this chapter, it is obvious that this approach of fuzzy-sets is in essence a special version of what is called 'possible-worlds-semantic' (pws) for intensional languages[14]. I will not give a detailed translation but only point out some resemblances (as well as differences) between Ludwig's and the pws-approach. In pws the interpretation of a first order language L consists of a set K of interpretations, whose members $I_D \in K$ stand in a 'family-relation' $\mathcal{R}$, which like Ludwig's relation between the $U \in N_\mu$ is usually reflexive, transitive and symmetric. Hence, one main difference is, that Ludwig strengthens the relation $\mathcal{R}$ by conditions 4) and 5) such that K resp. $N_\mu$ forms a 'uniform-structure'. In addition, one of the interpretations $I_p \in K$ can be distinguished as the 'intended' or 'ideal' interpretation such that the resulting 'imprecision-interpretation' of L consists of the quadruple $\langle K, \mathcal{R}, I_P, I_D \rangle$. At the same time, the absolute notion of truth of the Tarski-semantic becomes relativized to the notion of 'true in a certain interpretation'. At first glance, there seems nothing to correspond to this relativization of truth in Ludwig's approach, because he insists that the 'primitive' logic of basic propositions has not changed: "none of the axioms of the observational report with the imprecise-relations is only 'probably-true'". (Ludwig (1978), p. 56). Yet, we have to remember that the axioms of fuzzy-representation[13] are only true in the sense of a 'useful' description of facts if we choose for every relation $R_\mu$ an 'adequate' imprecision-set $U \in N_\mu$ to define a 'useful' imprecise-relation $R_\mu^U$. Contrary to what Ludwig seems to believe, this is plainly equivalent to the indicated relativization of truth in pws. Hence, in both cases only a relative definition of truth, a recursive definition relative to a choosen interpretation $I_D \in K$ resp. $U \in N_\mu$ can be given for every $s \in L$ in the usual way of Tarskian-semantics. However, the main point of interest in intensional languages and their pws lies in the circumstances (and there is a resemblance in Ludwig' approach too) that we can define a sentence-operator $\Box(s)$ such that: '$\Box(s)$' is true iff s is true in every interpretation $I_D \in K$, standing in the relation $\mathcal{R}$ to the intended 'ideal' interpretation $I_P \in K$. Hence, the sentence-operator '$\Box(s)$' turns out to be a modal one with the sense 's is necessary', which means by definition 'true in all possible worlds'. With the help of the necessity-operator more modal-operators like 'possibility' can be defined and indeed Ludwig defines in § 10 (1978) "Physikalische Möglichkeit, physikalische Wirklichkeit und Unentscheidbarkeit als Begriffe in einer PT" a battery of two times nine (!) modal concepts. I'll not discuss these now, except for one general remark: Modallogic and its semantic is a far from trivial extension of first-order-logic and its standard semantic! Indeed some people, like Frege, believe it to be a rather dubious one.

## VI. THE PHILOSOPHY OF 'IMPRECISION-SETS'

Now we are in a position to ask: What is the philosophy behind the 'technique' of concepts with 'imprecision-sets'?

(1) Does it mean that physical concepts in contrast to mathematical ones are indeed 'inexact', in the sense that we really don't know what their 'sense' and thereby their exact extension is?

(2) But if so [and much in Ludwig's books favours this view],does it mean that the 'inexactness' of physical concepts can be explained by the technique of 'imprecision-sets', taking families of mathematical exact concepts as pictures of reality which is in essence a 'possible-worlds-semantic'.

(3) Or is the technique of 'imprecision-sets' understood only as a <u>pragmatic device</u>, which is necessary to save physical theories from contradictions with 'reality', without presupposing that physical concepts are really inexact, (leaving the problem open to a 'meaning-theory')?

In trying to find an answer from Ludwig I'll quote some passages: "Oft ist der Sprachgebrauch so, daß man so tut, als ob das mathematische Objekt die exakte Situation sei, die aber durch die Feststellung - die Messung - realer Gegebenheiten nur "ungenau beobachtet" wird, d. h. wegen eines "Meßfehlers" nur unexakt bestimmt werden kann. Durch die Entwicklung der Physik sind wir aber gegenüber solchen Redewendungen skeptisch geworden: Die "an sich existierenden" aber nur ungenau festgestellten Tatsachen sind keine Basis für die Physik". ((1978), p. 49) "Die Theorie MT enthält keine <u>systematische</u> Anweisung, genau eine bestimmte Unschärfemenge auszuwählen. Das liegt natürlich daran, daß man ein physikalisches Problem nicht hat lösen können und deshalb eine Theorie MT wählt, die eine Realrelation durch eine Idealisierung $R_\mu$ darstellt, um dann diese "Idealisierung" nachträglich durch "Verschmierung" von $R_\mu$ mit Unschärfemengen U rückgängig zu machen"; and in connection with an example from geometry we read: "Die <u>kontinuierliche</u> Struktur von X in MT ist also <u>kein</u> Bild der Wirklichkeit, sondern ein Ausweg aus einer Unkenntnis durch Idealisierung. Genau diese Idealisierung müssen wir aber wieder rückgängig machen, indem wir bei Benutzung von MT als Bild das Prinzip der unscharfen Abbildung mit "geeigneten" Unschärfemengen anwenden", and one passage before: "Solche MT, die überhaupt keiner Unschärfemengen bedürfen, sind aber reine Utopie". [All quotations from (1978), P 51/52]

From all this, it seems to me, follows that the answer to question(3)is in the negative and to question (1) and (2) in the affirmative direction, or a bit more carefully stated, the answer seems to be the following:

Although we need - for pragmatical reasons - 'exact' mathematical theories for the description of 'reality', we must not mistake the 'ideal' mathematical structures for 'pictures of reality'. On the contrary, we have to be aware of the 'idealization' connected with mathematical theories, and this can be done only by a cancellation of the 'idealization' though the 'meta-mathematical' technique of imprecision sets, making the physical (not the mathematical) concepts 'inexact' because - the reason is simple - as empirical concepts abstracted from finite experience they are inexact! Hence, the point is really a semantic one: We know the sense and thereby the extension of a physical concept only up to a certain degree of imprecision in 'approximation' to reality. This is a consequent elaboration of the 'picture-conception' of meaning, namely that the sense of a physical concept-expression is only determined up to the 'exactness' of its measurement, that means the method of its verification.

## VII. THE CRITIQUE OF THE 'IMPRECISION-SET VIEW' OF CONCEPTS

Before I go one to analze this view and compare it with Frege's conception of concepts as functions in judgements, I should first make clear what I don't question:

1) I don't question that Ludwig's view is a sophisticated description of the leading praxis of what physicists do if they apply physical theories to 'reality' bridging the gap between theories and experimental results, yet without presupposing that physical concepts are 'inexact'.

2) I don't question that Ludwig's approach of 'uniform-structures' is a mathematical advanced technique adapted to the purposes of physics with its concepts mainly quantitative in character, which is far ahead of the simple technique of pws, used in modal logic resp. indexical semantics.

Rather my questions are in both respects more philosophical or, if you like, mor normative: First, what should physicists reasonably do, if they apply physical theories to experimental results and second, what should modal logic and especially its semantic look like, if it shall permit a rational discours about experimental results in the light of 'ideal' theories; a more rational discours than other more restricted positions like

Frege's judgement view will allow?

My first and most basic argument against the philosophical intentions of 'inexact' concepts is simply, that it stands the epistemological order between concepts and objects i.e. real objects, on its head, and instead asking for the reasons of the gap between physical concepts and their application to 'reality', the semantic of 'imprecision-sets' blames the physical concepts for the gap as if we could not know what their exact physical sense is. But why, I ask, should we blame the 'self-made' concepts instead of the experimental results for the gap as physicists have always done? Is there any reason to change the epistemological order between exact concepts and our incomplete (i.e. finite and restricted) knowledge of 'reality' manifested in the imprecision of quantitative results? (I'll return to this question in a moment.)

In every case, the argument that the reality is inexact and that this is the reason why we have to represent it by 'imprecision-sets' is, as a general argument, vastly over-extended, because it makes the same mistakes as the dual argument of the 'exact nature', supposing a certain 'ontology', by knowledge of the 'Ding an sich'. On the other hand, the weaker and more careful argument that we don't know what the exact structure of reality is, is no argument at all against 'exact' concept because otherwise, we would already know that we never could know what the exact structure of nature is; which claim is just as indefensible as the above ontological argument. Yet, it seems to me in regard to the quotations, that much of Ludwig's philosophy rests on an epistemological argument of the last kind, and hence, 'inprecision-sets' for physical concepts are for him a matter of principle.

But let us suppose for a moment that physical concepts have to be 'inexact' in the way defined. Can we grasp such concepts? Can we think with them? Can we understand what the 'inexactness' is or what it is an 'inexactness' of? There is the not only curious but rather serious fact, that the 'inexactness' of physical concepts is 'explained' by a set of mathematically 'exact' concepts, the uniform-structures $\underset{\sim}{N}_{\mu}$, by which we construct the set of 'imprecision-relations' $\tilde{R}_{\mu}$, as a substitute for the exact interpreted relation $R_{\mu}$ . Where does the 'physical' knowledge of imprecision come from, from a set-theoretical meta-theory? Why 'uniform structures'? Have we yet certain 'idealizations' in mind, and why not others? It seems to me that the 'inexactness' of physical concepts, should it really exist, is at best 'simulated' in the mathematical theory but not 'explained', as I understand that term. We don't see 'why' physical concepts should be irreducibly inexact, we don't see the physical reason; rather we claim that for pragmatical considerations physical facts have to be represented by 'sets of mathematically exact concepts'.

This brings me back to my first question: Are there any reasons to think that physical concepts, their sense and with it their extension, have to be 'inexact' in principle, no matter how this is 'simulated'? If we remember once more what the dilemma was which created the trouble, then the answer is simple: The application of exact physical concepts to 'reality', to real objects in concrete physical situations. However, it is important to understand that this answer leaves it completely open whether the reasons for the dilemma lie in our incomplete knowledge of real things or in our misconception of concepts as 'exact'. Now, one thing is trivially clear: without exact concepts the dilemma would not occur! But, is it sufficient for the dilemma that concepts are exact? Clearly not, because in mathematics as the 'exact' science, the dilemma doesn't occur. The reason is 'simple', in the case of mathematics we know the objects because we ourselves have created them. Hence, according to Mill's rule of the 'difference in circumstances', I conclude that the actual reason for the dilemma in the case of physics is our incomplete knowledge of concrete physical objects. So, why should we treat the symptoms by making concepts 'inexact' instead of treating the actual cause for the dilemma: our incomplete knowledge of real objects?

Now, I am aware, that one might have an argument, at first glance a striking one, in the bare existence of quantum-mechanics which in its 'Copenhagen-interpretation' denies that we can know certain facts more 'exactly' than up to a certain degree of 'imprecision', denoted by the Heisenberg uncertainty relation: $\Delta p \cdot \Delta q \geq \frac{\hbar}{2}$. Although Ludwig doesn't apply this argument, one must remember the special reasons for this abstinence: In this case, according to Ludwig, we know the physical reasons for the irreducibility of imprecision, the impossibility of preparing a micro-system such that its incommensurable observables, like position and momentum, can be measured 'together' more exactly than up to a certain degree of imprecision, determined by the Heisenberg uncertainty relation. Therefore, as Ludwig points out, the 'imprecision' in this case is one of preparation relative to registration and has nothing to do with the imprecision of measurement or imprecisions between theory and reality [L. this volume]. Hence, unless we accept quantum mechanics and its epistemic interpretation as the universally valid physical theory, it would not be legitimate to extend its epistemological features to the whole of physics. And yet, it seems to me, it is exactly, what Ludwig does: if he states that "Macrophysics is essentially based on these imprecisions of preparations which in principle cannot be avoided". The only difference between quantum mechanics and the general case is, as far as I can see, that in quantum mechanics we have already incorporated into the theory, what the smallest possible set of imprecision-relations is.

Underlying the opinion of a general inexactness of physical concepts in picturing reality is, I suspect, a muddle between grasping a concept as a constituent of our thinking and knowing the extension, that is the range of real objects falling under the concept, as the 'Bedeutung' of a physical-concept-expression: If the first, the grasping of the 'sense', is explained by the latter, the knowing of the extension, then there is no escape from the 'inexactness' of physical concepts, because we don't know what the exact structure of 'things in themselves' is.

What is the alternative? To give an answer, we have to remember Frege's conception of concepts as functions in judgements. According to this view, the 'sense' of any significant expression is "the way in which its 'Bedeutung' is given" - that means, for concept-expressions a certain combination of 'Merkmale', which can be predicated to certain objects, pairs of objects and so forth as their prospective properties and relations. These in turn form a part of the sphere of the sense of objects, their 'way of giveness". Hence, if we judge a thought to be true (or false), we judge nothing else than that the objects given in a certain way, expressed by the names in the sentence, have the properties and relations predicated by the 'Merkmale' of the concept. This is, what we really do in physics, if we assert certain propositions. Yet, it is very important to understand two points: First, that the sense of a concept-expression - in distinction to that of a name - is always an 'abstractum', which means, it is abstracted as a general feature from an indefinite set of properties and relations in which objects may be given, pointing out common aspects between different objects and disregarding all other differences. This process of abstraction is the intellectual operation which makes concepts general, regardless of how many objects in reality fall under them. Secondly, and equally important, is the circumstance, that only the sense of names - in distinction to that of concept-expressions - can be concrete, because only objects can be given in a definite and none the less contingent way - which means, they are given in a certain known way, one of indefinitely many different possible ways.

According to these considerations it cannot be expected - at least not rationally - that a real given thing can be completely determined in its properties, by a set of abstract concepts and their combination of 'Merkmale', because by the generality of concepts something is always lost. All that we can do reasonably is to 'interpret' ('deuten') the sphere of sense, by defining abstract concepts as functions in judgements such that there exist certain kinds of objects, pairs of such and the like, all given in a definite set of possible ways, iff they have the properties and relations predicated by the general concepts. As a first step we have to select and specify certain definite ways in which objects may be given, e.g. space and time in physics, and then to construct in the realm of these definite ways of giveness certain so called 'ideal' objects or models according to the defined concepts, concepts which establish connections between the different ways of giveness. In a second step we have then to reflect, if these constructions are possible. We have to judge the 'ideal' objects exist in the sense that they have the properties and relations predicated by the concepts as their 'way of giveness', and to that extent fall under the concept.

This is, how we proceed in mathematics and, in my view, should proceed in physics too! Yet, there is on fundamental difference: In mathematics, where we abstract our concepts from the material properties of objects, it is sufficient to construct the objects 'theoretically', whereas in physics with its material bodies, it is not sufficient to construct the objects only 'theoretically' (as in 'thought-experiments'). In doing physics we have to go a decisive step further: we have to construct the objects of a physical theory also in 'practice' - with concrete material bodies in real experiments, although the theoretical construction of ideal objects remains a necessary step for an understanding of a physical theory.

What distinguishes my position from that of Ludwig, is the acknowledgement of the construction of 'ideal' physical models as a necessary condition for the understanding of a physical theory, the physical (as opposed to the mathematical) meaning of its concepts and laws. This is a precondition for its rational application to real given things. Ludwig, on the other hand, immediately matches a range of real given things with the mathematical theory by the 'correspondence rules', without explicating the physical meaning of the concepts and laws constituing the physical theory by constructions of 'ideal' physical models. Therefore, he has no other choice to fill the gap between theory and experiments than by introducing 'sets of imprecision relations' as pictures of reality.

What is the solution to the dilemma? There is none because there is no dilemma, although there is a gap between concepts and reality. I agree that no construction in practice is perfect, measured against the 'ideal' objects of the theory. Yet, this is not astonishing given the role of concepts in science, namely achieving generality in thoughts by abstracting from contingent properties of real given things. The definition of highly abstract concepts like energy and momentum is the only way we can bring our thoughts to a timeless and spaceless form free of contingency. Hence, we cannot expect that the 'ideal' objects, which fall under the theory, are members of the realm of real given things; at best they stand in an approximation relation to them.

Nevertheless, I agree that the gap between theory and experiment has to be bridged. But one thing should be clear above all: There is, in my view, no direct comparison between experiments and theory; instead the relationship is split up into at least three parts: theory, its 'ideal' models and real experimental results. To analyze the relation between the first two is a purely theoretical task in respect to the truth of the theory. To minize or perhaps even close the gap between the last two is a theoretical and practical task, where modal considerations play an important role in respect to the completeness of the physical theory in regard to 'reality'. However, it is important to note, that in my view this task is in no sense the one-way-street of theoretically adapting the 'ideal' models to 'reality' by formulating new and 'better' theories (an impression one can have from Ludwig [1978]); but equally important is the process in the other direction, adapting the real objects in experiments to the 'ideal' constructions of the theory. This latter process is the only way we can learn systematically what properties and relations we may have neglected in our theories about the one physical world[15]. Therefore we need 'ideal' physical models and this 'presupposes' 'exact' physical concepts. Therefore, if Ludwig understands his 'technique of imprecision-sets' as only a pragmatic device to save physical theories (already judged as theoretically true) from immediate contradictions with experimental results, and also understands his modal concepts as only a mode for expressing how 'approximately useful' a physical theory is, I agree with him.

NOTES AND LITERATURE

(1) M. Dummett (1980), "Common Sense and Physics" ind 'Perception and Identity' - Essays presented to A. Ayer, ed. G.F. Mac Donald, Oxford 1980.

(2) G. Frege (1879), "Begriffsschrift" - 'eine der arithmetischen nachgebildete Formelsprache des reinen Denkens'.
(1884) "Die Grundlagen der Arithmetik" - 'eine logisch-mathematische Untersuchung über den Begriff der Zahl', Wiss. Buchg. - Darmstadt 1961.
(1891) "Funktion und Begriff"
(1892) "Über Sinn und Bedeutung"
(1892) "Über Begriff und Gegenstand" reprinted in "Funktion, Begriff, Bedeutung" ed. G. Patzig, Vandenhoeck & Ruprecht, Göttingen 1962.
(1893) "Grundgesetze der Arithmetik" - 'begriffsschriftlich abgeleitet', Wiss. Buchg. - Darmstadt 1962.
(1918) "Der Gedanke", "Die Verneinung", "Gedankengefüge" in 'Logische Untersuchungen', ed. G. Patzig, Vandenhoeck & Ruprecht, Göttingen 1966.
(posthum) "Nachgelassene Schriften und wiss. Briefwechsel", ed. H. Hermes, F. Kambartel, F. Kaulbach, Vol. I/II Felix Meiner, Hamburg 1969.

(3) H. Hertz (1894) "Die Prinzipien der Mechanik", Wiss.Buchg. - Darmstadt, 1963.

(4) E. Mach (1883) "Die Mechanik" - 'historisch-kritisch dargestellt' reprint of 9th edition, Leipzig 1933.
(1905) "Erkenntnis und Irrtum", 3th edition, Ambrosius Barth-Leipzig 1917.

(5) G. Ludwig (1970) "Deutung des Begriffs 'physikalische Theorie' und axiomatische Grundlegung der Hilbert-Raumstruktur der Quantenmechanik durch Hauptsätze des Messens", Springer-Lecture Notes in Physics 4, Berlin, Heidelberg, N.Y.

(1978) "Die Grundstrukturen einer physikalischen Theorie", Springer-Hochschultext.
(1979) "Einführung in die Grundlagen der Theoretischen Physik", Vol. I-IV, Vieweg-Braunschweig.

(6) In this chapter, I give a short summary of "my" understanding of Frege's conception of concepts. My presentation is insofar "dogmatic", as I neither give an explicit philological documentation of my view by Frege's writings, nor I try to defend my view against deviating interpretations. Both is the task of another paper, nevertheless, I hope, my view is correct. My understanding of Frege has most profited from the following three philosophers and their works.
Ch. Thiel (1965) "Sinn und Bedeutung in der Logik Gottlob Freges" Verlag A. Hain - Meisenheim.
M. Dummett (1973) "Frege - Philosophy of language" Duchworth - London.
H.D. Sluga (1980) "Gottlob Frege", Routledge & Kegan Paul, London.

(7) A. Trendelenburg (1840) "Logische Untersuchungen" S. Hirzel, Leipzig, 3. Edition (1870).

(8) H. Lotze (1874) "Logik - drei Bücher vom Denken, vom Untersuchen und vom Erkennen" ed. G. Misch, Felix-Meiner-Hamburg (1912).

(9) Moreover, we have here the rare but interesting case, that the formulation of a theory contains - beside mentioning of the universal 'Kepler-constant' - a proper name, that of the sun, and hence refers in its first law to a contingent object, called 'sun'. Insofar is the concept of Kepler's theory not of the most general form, possible. For an interesting kinematic theory and Newton's dynamic theory with gravitional forces, see:
E. Scheibe (1973) "Die Erklärung der Keplerschen Gesetze durch Newtons Gravitationsgesetz" in 'Einheit und Vielheit' Festschrift für C.F. von Weizsäcker, Vandenhoeck & Ruprecht, Göttingen.

(10) Evans & Mc Dowell (1976) "Truth and Meaning" Oxford-Univ. Press

(11) M. Hesse (1961) "Forces and Fields" - 'The concept of action at a distance in the history of physics', Nelson and Sons, London.

(12) L. Boltzmann (1892) "On the methods of Theoretical Physics" in L. Boltzmann, 'Theoretical Physics and Philosophical Problems', ed. B. McGuiness, Vienna circle collection Vol. 5, Reidel, Dordrecht-Holland, (1974).

(13) Here I depart from the editor's recommendation to translate 'Abbildungsaxiome' by 'observational report' and 'Bildmengen/relationen' by 'interpreted sets/relations'.

(14) D. Gallin (1975) "Intensional and Higher-Order Modal-Logic", North-Holland Publ.Comp., Amsterdam, Oxford, New York.

(15) What the method of such learning is, how we proceed rationally in fitting our physical theories to experiments, 'et vice versa', I have already explained some years ago in:
U. Majer (1975) "Paradigmatische Erklärungen und die Kontinuität der Wissenschaften" in 'Logik, Ethik, Theorie der Geisteswissenschaften', XI. Deutscher Kongreß für Philosophie, Felix Meiner Hamburg 1977.

The point was, that we can correct our theories as well as experiments only by expanding them taking new 'Merkmale' resp. properties or new functional dependencies between them into account.

# DIFFERENCES IN INDIVIDUATION AND VAGUENESS

W. Grafe

Philosophisches Seminar der Universität
Nikolausberger Weg 9c
3400 Göttingen
Federal Republic of Germany

## I. EPISTEMOLOGICAL SUGGESTIONS

From an epistemological view, classifying a statement as 'vague'[0] means to judge the statement in question to be a mixture from partial knowledge and partial ignorance. Accordingly it seems desirable to describe the boundary between knowledge and ignorance hidden in the vague statement.

Ludwig discusses [1] vagueness in physics, especially vagueness in measuring statements. The example he uses is 'measurement of Euclidean distance', i.e. the meaning of statements which are often written as "$d(x,y) = \alpha \pm \varepsilon$", where vagueness is expressed by "$\pm\varepsilon$" indicating the so-called "error of measurement". Ludwig maintains that physicists have come to refrain from supposing that physical objects have exact properties which cannot be measured exactly (but only within the indicated 'error of measurement'). The argument substantiating this attitude is obviously that the ascription of precise properties to physical objects is beyond the reach of physical theorizing. But what is the alternative? At first it seems that Ludwig would accept a rivalizing supposition to the effect that physical objects do have 'vague' physical properties. This indeed would be a rather puzzling point of view, not in accord with the course adopted in the beginning. Vagueness of statements there was taken to represent some deficit of knowledge but not a knowledge of a special kind, viz. that some objects does have a 'vague' property or that $\underline{n}$ objects (in a certain ordering) stand in a vague' $\underline{n}$-place relation. Now, nothing in Ludwig's formal treatment of 'vagueness' must be understood to imply this 'puzzling view', but on the other hand nothing in Ludwig's exposition is apt to exclude it. Are we then to take the well defined Euclidean distance function $d(x,y)$, which is as precise a term as you could imagine and pervert it into a 'vague three place predicate' when applied to empirical objects? And what does such application consist in? Or are we to hold the 'naive view' that physical objects x,y are (at one instant of time) separated by a precise distance, which simply cannot be measured exactly? My way out of this dilemma is to give a rather detailed description of how vagueness gets in, without assuming any predicates to be 'vague'.[2] This description amounts to pointing out different means of individuation of objects on different levels of argument and reducing statements of vagueness to statements of difference in individuation.

## II. LUDWIG'S EXEMPLIFICATION OF 'VAGUENESS'

Let me indicate what I hold to be the main features of Ludwig's account of vagueness in the case of measuring d(x,y) and then scetch my point of departure.

To restate the example, we need a rough restatement of the MTA-component of Ludwig's structuring PT of the intuitive concept of "physical theory":

MT is a mathematical theory, containing some (rather weak) set-theory as a subtheory and, in addition to this, 'special axioms'. A suitable two-valued logic is assumed to be incorporated in the language of MT.

G is to be conceived as a kind of 'open collection' (not to be taken as a well defined set), which comprises among other things 'pre-theories' and labelled empirical objects $a_i$ ("genormte Realtexte").

"(——)" is a set of correspondence rules which

1. single out terms $Q_j$ of MT;
2. single out some relations $R_n$ of MT ($R_n$ has $K_n$ places);
3. establish rules how to
   a) state sentences of the form $\ulcorner a_i \in Q_j \urcorner$
   b) state sentences built up from the relations $R_n$ given in 2. taking as arguments the $a_i$ and perhaps specific real numbers and negations of such sentences.

In order to cope with vagueness, 3.b) has to be modified by replacing the $R_n$ by weaker $\tilde{R}_n$.

MTA may then be understood as the extension of MT resulting from the application of the correspondence rules (——) to the $a_i$ in G. To restate the example, MT is supposed to contain metrical Euclidean geometry with the distance function d(x,y) defined as a map: $\mathbb{R}^3 \times \mathbb{R}^3 \rightarrow \mathbb{R}_{0+}$ which satisfies the usual equation[3].

The $a_i$ in G are supposed to label marked places in physical space, the only $Q_j$ needed is $\mathbb{R}^3$, the only $R_n$ is $R(x,y,\alpha)$, defined as $d(x,y) = \alpha$. Thus the statement obtained by 3. of (——) are:

a) $a_i \in \mathbb{R}^3$;[3]

b) the measurement-results from G reported in terms of $R(x,y,\alpha)$.

To avoid inconsistency of MTA only due to the limited accuracy of measurements, the statements obtained by 3.b) of (——) have to be replaced by analogous statements, in which $R(x,y,\alpha)$ has to be replaced by $\tilde{R}(x,y,\alpha)$, which is defined as follows: First define some imprecision-set U for some fixed real number $\varepsilon > 0$ by $U = U_\varepsilon = \{\langle\alpha,\beta\rangle \mid \alpha,\beta \in \mathbb{R} \wedge |\alpha-\beta| \leq \varepsilon\}$ . Then $\tilde{R}$ is defined by

$$R(x,y,\alpha) \Leftrightarrow \bigvee_\beta[\beta \in \mathbb{R}_+ \wedge R(x,y,\beta) \wedge \langle\beta,\alpha\rangle \in U] \quad \text{for } x,y \in \mathbb{R}^3 \text{ and } \alpha \in \mathbb{R}_+.$$

Here I conclude my summary of Ludwig's account, because the point I am interested in does not depend on the further refinement of the method of imprecision-sets. Two points seem to me indisputable and I think they are held by Ludwig too:

1. $\tilde{R}$ is a term of MT, i.e. a purely mathematical term just as R, because it is explicitely defined in MT without any reference to G or (——). So $\tilde{R}$ is as exact or precise a relation as R.
2. The claim that $\tilde{R}(x,y,\alpha)$ is needed to replace $d(x,y) = \alpha$ in the statement according to 3.b) of (——) implies that not all of the $a_i$ are uniquely represented in MTA by single elements of $\mathbb{R}^3$ (because $d(x,y)$ is a function, taking a unique value for uniquely identified arguments).

These are the features in Ludwig's presentation of the example "measurement of $d(x,y)$", which I found in need of analysis. The observations 1., 2. above do not decide anything with respect to the 'puzzling view' sketched in section (I), because from Ludwig's treatment it is not clear what is meant by the application of $d(x,y)$ to empirical objects. The source of the trouble is that the statements obtained by 3.a) of (——), which assert the $a_i$ to be elements of $\mathbb{R}^3$ are unacceptable as they stand. Instead of statements of the form $a_i \in \mathbb{R}^3$ , we should concentrate on statements of the form $\varphi(a_i) \in \mathbb{R}^3$, where $\varphi(a_i)$ denotes some mathematical objects, while keeing in mind that in order to cope with 'vagueness', we shall have to consider not one single map $\varphi$, but some set of such maps. Now Ludwig may well maintain that a suitable variety of such maps is assumed to be supplied by (——). My point is merely that we do not understand what is meant by "MTA contains vague statements" without information about what such mappings look like. Thus, in a sense, the rest of this paper is concerned with the task of understanding what is meant by the statement 3.a) of (——), or better: by what sort of statements they should be replaced.

## III.1. CONCEIVING 'VAGUENESS' AS RESULTING FROM DIFFERENCES IN INDIVIDUATION

Before specifying the example, I must list some properties of what I propose to call a "pragmatically controlled (data-) language", conceived as a first order language. The next step then is to conceive a set of data formulated in such a language as a first order theory, say: the 'data-theory'. As we may suppose that such a (finite) set of data has a truth evaluation (is truth-value definite for any datum in the set) on empirical grounds, we may easily fix even the set theoretic interpretation of this data-set. The next, and I hope, the decisive step consists in confronting the doubly interpreted data-set (empirically and set-theoretically interpreted) with the corresponding "idealized measurement structure". This structure is well known from the literature and I take it to be expressed as a first order theory plus set theoretic interpretation (which I choose to be model-theoretic). The vagueness statement then is expected to result from the combination of the two clearly different set-theoretical interpretations. To summarize not only the procedure but also the thesis: A prominent property of a "pragmatically controlled (data-) language" will be that empirical objects, which figure as the denotata of the individual constants of the language, are identified operationally, and consequently by finitary means. (The truth-value decisions of the respective sentences are also approached according to finitary methods.) This, in my understanding of theorizing in physics, is in sharp contrast to any means used by some physical theory in describing (individuating) its objects. Restated: theoretical characterization of objects is non-finitary, e.g. by means of a theory admitting only infinite models (in the model-theoretic sense). Accordingly we should not be astonished, if the difference between theoretical and operational individuation is expressed as some 'vagueness'. (There is no reason to suppose such vagueness of physical assertions as representing a problem 'not yet' solved by physical theory. Of course the possibility of varying the sort or degree of vagueness will depend on the physical knowledge available at the time. But it seems quite plain that such vagueness is rather a consequence of theory-construction in physics, viz. that the structure explained are 'idealized structures' only, which results from the use of mathematical methods.) I do not claim that the contrast "individuation by finitary means" vs. "individuation by non-finitary means" is the only noteworthy thing about "operational" and "theoretical" individuation; but I claim that it suffices as a first attempt to describe how 'vagueness' gets in, as far as our example is concerned.

## III.2. A DATA LANGUAGE FOR BASIC MEASUREMENTS

First I shall state some conditions (1) on any pragmatically controlled language of a set of data and secondly state conditions (2) on a data-laguage for basic measurements of lengths by rods. This is a considerable simplification of Ludwig's example (Euclidean geometry), but I think it will serve the purpose of demonstration just as well.

(1) L is a first order language with identity and with only finitely many individual and predicate constants. For each indi-

vidual constant ⌜$a_i$⌝ the object denoted is operationally identified, and this object is denoted by no other individual constant. For any atomic sentence of L the truth-value can be established according to a set of accepted methods. (These methods may, but need not all be operational.) A set D of data in L is a finite set of quantifier-free sentences in L, including for any individual constant ⌜$a_i$⌝, that occurs in some member of D, the equation ⌜$a_i = a_i$⌝. Moreover D is consistent and the sentences of D have been decided according to the accepted methods.[4)]

(2) The predicates for length-comparison by rods include two two-place predicate constants "R" and "G", which shall be understood as follows: "R" denotes the operational "greater than" and "G" the operational relation "is equal in length to". The accepted methods for assigning truth-values shall include: for no $a_i$ is ⌜$Ra_i\ a_i$⌝ or ⌜$Ga_i\ a_i$⌝ true. The language may or may not contain a further three-place predicate constant "Add" denoting "addition of lengths".

Suppose, the length comparisons lead to results of the kind we expect of such comparisons from earlier experience. Then we may define a finite weak order if D is complete in the sense, that <u>all</u> sentences of the forms ⌜$Ra_i\ a_j$⌝, ⌜$Ga_i\ a_j$⌝, ¬⌜$Ra_i\ a_j$⌝, ¬⌜$Ga_i\ a_j$⌝ have been decided as true or false respectively and D is the set of sentences thus decided as true.

An equivalence relation may be defined by ⌜$Exy \Leftrightarrow Gxy \vee x=y$⌝, and the reflexive order relation may be defined by ⌜$Sxy \Leftrightarrow Exy \vee Rxy$⌝, thus expanding the data-language by two two-place predicate constants "E" and "S". As domain for the finite weak order $\mathcal{E}$ we choose (1-1) a denotation for any ⌜$a_i$⌝ occuring in some sentence of D. $\mathcal{E}$ is then determined uniquely up to isomorphism as a finite relational structure. Hence the set-theoretical interpretation of D is unique and so is the truth-evaluation by the operational identification of the $a_i$ and the operationally decided predicates "R" and "G".

Keeping in mind the goal of comparing $\mathcal{E}$ with some idealized measurement structure, it seems fair enough to expand $\mathcal{E}$ by adding "Add" to the relations "R", "G", "E", "S", taking its values in accordance with the operational results. But note that the set of $a_i$'s is not enlarged by allowing the operationally interpreted "Add". Of course we may assume D', resulting from D by adding the atomic "Add"-sentences or their negations respectively, to be complete in the quantifier-free sublanguage of L. So D' defines uniquely a finite relational structure $\mathcal{E}'$, which is simply the expansion of $\mathcal{E}$ by "Add".

## III.3. AN IDEALIZED MEASUREMENT STRUCTURE

As an idealized structure for basic measurements of lengths we consider the 'positive closed extensive structure' of Krantz et al. (1971). It differs from the finite weak order $E'$ mainly in being closed with respect to "length-addition".

Krantz's idealized addition operation "o" exhibits the Archimedian property. It can be proven (cf. e.g. op.cit., pp. 74 f) that there is a homomorphism into the real numbers, which is unique up to the choice of some positive scale factor. For nearly all finite structures like $E'$, this does not hold[5]. As the idealized structure is closed with respect to length-addition and hence infinite, there is no sense in taking (even some of) its objects to be operationally identified.

I shall now write down as first order formulae the axioms for the 'idealized measurement structure', taken from Krantz et al. (op. cit., p. 73), where $\succsim$ and o are taken as basic predicates.

[0.1] $\bigwedge xy[x \succ y \Leftrightarrow x \succsim y \wedge \neg y \succsim x]$; Def. of "$\succ$" in terms of "$\succsim$"

[0.2] $\bigwedge xy[x \sim y \Leftrightarrow x \succsim y \wedge y \succsim x]$; Def. of "$\sim$" in terms of "$\succsim$"

[1.1] $\bigwedge xyz[x \succsim y \wedge y \succsim z \Rightarrow x \succsim z]$; Transitivity of "$\succsim$"

[1.2] $\bigwedge xy[x \succsim y \vee y \succsim x]$; "$\succsim$" is connected

[2] $\bigwedge xyz[x \circ (y \circ z) \sim (x \circ y) \circ z]$; "o" is weakly associative

[3] $\bigwedge xyz[x \succsim y \Leftrightarrow x \circ z \succsim y \circ z \Leftrightarrow z \circ x \succsim z \circ y]$; monotonicity

[4.1] $\bigwedge x[1x = x \wedge \bigwedge_n (n+1)x = nx \circ x]$; "nx" recursively defined

[4.2] $\bigwedge xy[x \succ y \Rightarrow \bigwedge_{zu} \bigvee_n (nx \circ z \succsim ny \circ u)]$; 'Archimedian property"

[5] $\bigwedge xy[x \circ y \succ x]$; positivity.

[0.2] implies what is sometimes called the "weak antisymmetry of $\succsim$" and so any model of [0.2] $\wedge$ [1.1] $\wedge$ [1.2] is a weak order. Of course [1.2] implies the reflexivity of "$\succsim$". Infinity comes in by axioms [4.1] - [5].

"$\succsim$" is the idealized counterpart of the reflexive order-relation "S" defined in III.2. "$\succ$" corresponds to "R", "$\sim$" to the equivalence relation "E" and "o" corresponds to "Add". As noted above it is via the properties of "o" that infinity comes in. This is due to the metalogical requirement, that any model of the axioms has to be closed with respect to the composition "o". Nothing like this is required for model-theoretic interpretations of predicate-constants. This is why for the empirical (data-)language, we had to choose "Add" as a predicate-constant rather than an operation symbol[6].

Axiom [4.2] (together with [4.1] asserts the existence of lengths transcending any given length. This is what I call 'idealized rod production', having no counterpart in an empirical domain.

## III.4. THE STRUCTURE $\mathcal{U}$

The set of (first order) logical consequences of the axioms [0.1] - [5] I call $T^{+}_{\lessgtr}$. The consistency of $T^{+}_{\lessgtr}$ is assumed. Accordingly, $T^{+}_{\lessgtr}$ is assumed to have a model, and as $T^{+}_{\lessgtr}$ only admits infinite models, $T^{+}_{\lessgtr}$ has a countable model. We single out one such countable model and refer to it as "$\mathcal{U}$", $\mathcal{U}$ is a set-theoretical entity of the same sort as $\mathcal{E}'$, our uniquely determined model of the set of data D'. We may now choose an object from $\mathcal{E}'$ (i.e. some set-theoretic picture of some $a_i$) to be the unit element. The unit of $\mathcal{E}'$ of course must be uniquely pictured in $\mathcal{U}$. What about the pictures of the remaining $a_i$ in $\mathcal{E}'$? The sentences of D' and the choice of the unit in the empirical domain and thus in $\mathcal{E}'$ do not in the least determine a unique embedding of $\mathcal{E}'$ in $\mathcal{U}$. The only exception is: $\mathcal{E}'$, and thus the empirical domain, contains only objects which belong to some "initial segment of a standard series". A standard series is a sequence of objects such that its n-th element equals in length n times the unit length. I refer to this exception as the "ruler-case".

## IV. CONCLUSION

In order to get a description of how mathematics is applied to empirical objects in basic measurement, I gave a description of basic measurement which is sumarized in the scheme below.

(a) Operationally identified $a_i$ and operationally decided predicates "R", "G" and "Add";

↓ determine empirically

(b) the set of data D', formalized in the quantifier-free sublanguage of the indicated language L

↓ determines logically, up to isomorphism,

(c) the finite weak order $\mathcal{E}'$, containing a value for "Add" too; set-theoretic entity of the type 'relational structure'

↓ isomorphic embedding, not unique, except in the ruler-case

(d) idealized measurement structure $\mathcal{U}$, which is a model of $T^{+}_{\lessgtr}$ (of § III.3.), a countably infinite relational structure

↓ homomorphic embedding, unique up to a scale factor

(e) the set of positive real numbers $\mathbb{R}_+$ with $\leq$, $=$ and $+$, $\cdot$.

My thesis is now a comment on the scheme above:

(i) It is an empirical question, whether the operational verification of the sentences of D' leads to results of the kind that we expect, i.e. whether D' turns out to be consistent, determines a weak order, and does not conflict with the axioms for "o" in its

"Add"-sentences. But if this is the case, $E'$ is determined uniquely by D' by purely logical and set-theoretical means.

So far, there is no chance for 'vagueness' to get in. Though $E'$ is a set-theoretical construct, it is only a set-theoretical paraphrase of an empirically determined truth-evaluation. The unique truth-evaluation is just the only thing, which is common to D', conceived as operationally interpreted, and to D', conceived as the set of quantifier-free sentences true in $E'$. This double interpretation is the only link between empirical and set-theoretical semantics in the example. But 'vagueness' did not appear on the stage of (a) - (c).

(ii) Obviously the source of 'vagueness' is located in the set $I(E',U)$ of isomorphic embeddings $\varphi$ of $E'$ into $U$. This is, because the remaining part of the construction is the homomorphic embedding $\underline{h}$ of $U$ in $\langle \mathbb{R}_+, <, =, +, \cdot \rangle$, uniquely determined up to a scale factor.

(iii) Now we are in the situation to state some features of 'vagueness' and how it gets in. First, we might conclude, that none (!) of the sentences considered is vague, because to any sentence a set-theoretic truth-condition is assigned, which would e.g. not be the case, if the methods from box (a) did not lead to yes-no-decisions in all cases. Secondly, what is my answer to the question of what the sentences obtained by Ludwig's (—) 3.a) are apt to mean?

Well then, take the whole, consisting of boxes (a)-(e) and their interrelations, to be a part of PT in Ludwig's sense. Define mappings, that transport any $a_i$ to $\mathbb{R}_+$. Part of the definition of any such mapping is some $\varphi \in I(E',U)$. We know the set $I(E',U)$ to contain not one, but 'many' elements (except in the ruler case). The elements of $U$ are pictured in $\mathbb{R}_+$ by $\underline{h}$ uniquely (unit given). Thus the ambiguous representation of our empirical objects $a_i$ within $\mathbb{R}_+$ is determined by their ambiguous representation in $U$.

What sort of 'vagueness' is characterized by the set $I(E',U)$? First, it is not 'error of measurement', since we assumed the measurements to be as precise as can be expected at all, namely to lead to definite yes-no-decisions. Secondly, it would be systematically misleading to say that "we do not know, which mathematical statements should be 'equated' with the empirical statements". There is nothing to know here. What is to be explained, is the way in which empirical statements are correlated with mathematical ones. That this correlation is not unique, and in principle cannot be unique, is simply a consequence of theory construction. The (—) 3.a) statements then have to be replaced by statements of the form "$\psi(a_i) \in X$", where both, $\psi(a_i)$ and X are terms of MT. The story to be told is that application of MT terms to empirical objects does not take place.

(iv) Let me close the paper with some remarks concerning 'individuation'. The fact that the embedding of $E'$ into $U$ is not uniquely determined, is drawn from the enormous differences between the way

$\mathcal{E}'$ was determined as opposed to the way $\mathcal{U}$ was. But we should refrain from stating this difference in terms of 'operational' vs. 'axiomatic'. The reason is, that on the stage of box (c) only one characteristic of 'operational truth-value-determination' is left: $\mathcal{E}'$ is characterized by finitary means, exemplifying "finite individuation". Only this information is transported to $\langle \mathbb{R}_+,<,=,+,\rangle$, nothing else can be confronted with the set-theoretic interpretation $\mathcal{U}$ of $T^{+}_{<}$. Again, the term 'axiomatic' does not characterize $\mathcal{U}$ as different from $\mathcal{E}'$, since we used a finite set of axioms to characterize $\mathcal{E}'$ uniquely up to isomorphism. This is even more than can be expected of an axiomatic characterization of $\mathcal{U}$, as $T^{+}_{<}$ is a theory of first order admitting infinite models. I think the difference is well described by stating, that operational individuation is by finitary means, which feature is preserved on the stage of $\mathcal{E}'$, while the individuation of objects of $\mathcal{U}$ is by non-finitary means, i.e. by means of an operation ("o") which admits only infinite models.

I do believe that this contrast between $\mathcal{E}'$ and $\mathcal{U}$ is not an accident, but caused by the different purposes they served. The purpose of $T^{+}_{<}$ and, accordingly, of $\mathcal{U}$ is to define a certain concept of measurement, that is general enough to subsume all special cases of sets of data like D', and also definite enough to imply the uniqueness of the map into the positive reals.

(v) The last thing I have to explain is, why I call the above mentioned contrast one of 'individuation'. Could one not as well call it a contrast of, say, 'types of predication'?

I do not think so, if one accepts a set-theoretical frame for describing theories, which is common to Ludwig's approach and to model theoretic treatment: The predicates or functions are set-theoretical constructs over the set of 'objects'. Thus, no predicate or function can be well defined unless the base sets are defined, but not vice versa. Thus the answer depends on a methodological decision. But what I hold to be independent of any such decision, is that the contrast of finitary vs. non-finitary means produces the so-called 'vagueness' of statements of basic measurements. Such 'vagueness' is understood to be unavoidable, as far as physics is designed to compare empirical facts with mathematical entities produced by 'non-finitary' methods.

From the epistemological point of view: It turned out that there are no 'vague statements', but contrary to common belief measurement statements are not applications of mathematical terms to empirical objects. There is no such application, and no such application is needed to describe measurement. There is only a correlation of empirical statements with mathematical statements, which is not unique. The (limited) arbitrariness exhibited shows the 'naive view' as well as the 'puzzling view' mentioned in section (I) not only to be badly substantiated, but from the outset unpromising as an attempt to characterize the way, measurement works.

FOOTNOTES

0) single quotation marks ',' are the 'ironic ones), the double ones "," are used to cite (interpreted) expressions in the usual way; Quine's quasi-quotation signs ⌜,⌝ are also occasionally used. No strict treatment of quotation is intended, but hints may seem useful and thus are given.
1) my references to Ludwig's treatment of the example "measurement of d(x,y)" as an example of 'vagueness of measurement' are to: Ludwig (1970), II § 6; (1978a), § 6; (1978b), II §§ 1-2, and III § 5.
2) Thus I reject the description, that 'vagueness of measurement-statements' is resulting from the use of vague predicates in the sense in which "fish" might have been vague once with respect to whales, or "bird" with respect to bats. In fact my thesis might well be understood to imply that 'vagueness' is a misnomer in the cases considered, for if we know anything relevant which is knowable at all, the phenomenon called "vagueness" will still be present.
3) I do not see any relevant modification in taking d(x,y) not to be a function of the sort described, but a function $d : X \times X \rightarrow \mathbb{R}_{0+}$, where X is some set constructed from a base set Y that is different from $\mathbb{R}$ but satisfies the same axioms. Thus I take X to be $\mathbb{R}^3$, as Ludwig does in (1978b), p. 10.
4) Note first, that the equation ⌜$a_i=a_i$⌝ are empirical assertions: they state, that the denotations of the ⌜$a_i$⌝ do not change (during the measurements recorded in D) in the properties expressed by the predicate constants occuring in some member of D. Note further that the condition of the consistency of D is independent of the condition that all members of D are decided as true according to the accepted methods, and so that both are needed.
5) the exception is simply what I term below in section III.4. the 'ruler-case'
6) Similarly stated in Krantz et al. (1971), pp. 81f.

LITERATURE

C.C. Chang, H.J. Keisler, 1973, "Model Theory", Amsterdam.
D.H. Krantz, R.D. Luce, P. Suppes, A. Tversky, 1971, "Foundation of Measurement", Vol. I, New York.
G. Ludwig, 1970, "Deutung des Begriffs 'physikalische Theorie' und axiomatische Grundlegung der Hilbertraumstruktur der Quantenmechanik durch Hauptsätze des Messens", Berlin.
G. Ludwig, 1978a, "Die Grundstrukturen einer physikalischen Theorie", Berlin
G. Ludwig, 1978b, "Einführung in die Grundlagen der Theoretischen Physik", Vol. I, 2nd edition, Braunschweig.

# A GENERAL SCHEME FOR INTERTHEORETIC APPROXIMATION[1)]

C.U. Moulines

Instituto de Investigaciones Filosóficas
Torre de Humanidades, 40 Piso
Ciudad Universitaria
Mexico 20, D. F.

## INTRODUCTION

This essay has a twofold purpose: first, to show that it is possible to introduce, in a natural way, a general structural concept of approximation between conceptually different theories; second, to show its viability by testing it against a particular example of an intertheoretic relationship, namely, the relationship between Kepler's laws and Newton's theory of gravitation. It is formally shown that this particular case fits into the general frame for approximation developed in the first part of the essay.

The attainment of the goal proposed has not only an interest in itself, but is of relevance for two more general issues within recent philosophy of science: the question of incommensurability of theories separated by a conceptual change, and the problem of building a systematic typology of intertheoretic relations. In this essay, it is not argued that our goal bears upon both issues; that this is so, should appear quite clearly once the results above-mentioned are obtained. In fact, if theories that apparently have a different conceptual structure can nevertheless be said to approximate each other in a well explicated sense of "approximation", then this would shed a new light on the question of incommensurability. And if a classification of different shorts of possible intertheoretic approximations can be provided in a natural way, then it is clear that any future typology of intertheoretic relations should take account of this fact.

In order to fulfill this double task, I have made extensive use of the ideas and results of three different approaches of presentday philosophy of science; they are associated mainly with the names of Sneed, Ludwig, and Scheibe, respectively. In trying to combine these three approaches, I am conscious of the danger of misinterpreting or "forcing" the three of them. But I am ready to take this risk. I think that the problem which is of our concern here is difficult enough to warrant the attempt to look for help coming from different sides. Besides, the three approaches mentioned show, in spite of important differences, a certain degree of methodological "family resemblance". I am thinking mainly of a common basic attitude in the reconstruction of physical theories that tends more to the consideration of global structures instead of single statements and to an extensive use of the tools of set theory and model theory for this purpose. It is this possibility of convergence on some fundamental matters that makes the attempt at combination appear as promising.

More concretely, what I will take from the authors mentioned is this. From Sneed's metatheory, besides the general jargon, the explication of the reduction relation; from Ludwig's approach, the idea of using uniformities to deal with any kind of approximative relationship; from Scheibe's work, the reconstruction of the relationship between Kepler's and Newton's theory.

Now, it is simply a fact that physical theories do not ever work exactly, but only with a certain degree of inaccuracy. This is true of the application of single theories as well as of many of the relationships that may hold between given (different) theories. This fact is known by everbody, but only recently philosophers of science have begun to take it seriously. Most of them appear to have taken it as a secondary, methodologically rather uninteresting aspect of physical science. As far as I know, Professor Ludwig was the first to perceive the approximative features of physics as crucial and non-eliminable even in principle. He saw them as an essential phenomenon both of the application of every single theory (I call this aspect "intratheoretical approximation") <u>and</u> of the relationships between theories ("intertheoretical approximation"). Many paradigmatic examples of relations between theories which are of great relevance to general philosophy of science are clearly cases of approximations - or, as one could say in more colloquial terms, of "blurring" of relations between theories. Well-known examples are: the relation between Kepler's theory of planetary motion and Newton's theory of gravitation; the relation between Newton's general mechanics and relativistic mechanics or quantum mechanics; the relation between (a part of) phenomenological thermodynamics and statistical machanics; the relation between geometrical optics and undulatory theory.

Further, it is also very important for the construction of a typology of intertheoretic relations to have a formal, general conceptual apparatus allowing the systematic treatment of all cases of approximation. A unified treatment of different kinds of intertheoretic approximation is a crucial step in the process towards a systematic description of intertheoretic relations of any kind. Accordingly, our basic aim is this: To develop a general notion of intertheoretic approximation that should be applicable to different cases of "blurred" intertheoretic relations. The general concept should have a sufficient degree of flexibility in order to cover different specific kinds of approximation between theories. This means that the general requirements put on the notion of approximation should not be too strong. Further, we would like to exemplify the functioning of the general notion. That is, we want to apply it to a particular case. The example I have chosen is the relationship between Kepler's and Newton's theories of planetary motion. This particular case has already been reconstructed by E. Scheibe in his (1973) - quite independently of any general metatheory. Scheibe's reconstruction is the most detailed I know of a particular case of intertheoretic relation. On the other hand, it is neither too complicated nor too esoteric. This is why it seems to be a most adequate on to test any general approach to intertheoretic approximation.

The strategy to be followed in order to attain our goal consists of the following steps.

1. To introduce the general apparatus for approximation (mainly, the notion of 'imprecision-set' and of model-theoretic 'neighborhood').
2. To provide a general structuralistic scheme for the relations of (intertheoretic) approximation and of almost exact (intertheoretic) approximation.
3. To define a particular kind of approximative reduction.
4. To reformulate (within the Sneedian structuralistic framework) Scheibe's analysis of the Kepler-Newton case. This implies, among other things, that Scheibe's conditions have to be identified and restated.
5. To prove that Scheibe's conditions for the Kepler-Newton case follow from the general concept of approximative reduction when some concrete parameters are inserted in the general definition.

Steps 1., 2. and 3. are intended to be universally valid for any study of intertheoretic approximations. Steps 4. and 5. should implicitly reveal the patterns to be followed in the analysis of other particular cases of intertheoretic approximations.

The metatheoretical frame of our task is (Sneedian) structuralism. I will assume that the reader is already acquainted with the basic elements of this approach. If this is not the case, he is referred back to the already voluminous literature of and on structuralistic methodology. He can find the most up-to-date overview of it in Stegmüller (1979), especially § 4. Our analysis will rest on the notions of the "second phase" in the development of structuralism, i.e. on the view of theories as tree-like nets.

Without going into the details, it might appear convenient shortly to summarize the central structuralistic notions that will be used here with some simplifications.

- It is assumed that the basic structures of empirical theories can be best identified by means of set-theoretic predicates _à la_ Suppes.
- It is further assumed that we can make a two-level, relative distinction among the concepts of a theory between T-theoretical and T-non-theoretical concepts.
- If T is a theory (in an intuitive, non-reconstructed sense), then M(T) is the set of its models, i.e. of those structures that satisfy the structural as well as the proper axioms of T.
- $M_p(T)$ is the set of all potential models of T, i.e. of those structures that satisfy the structural axioms of T (= the 'conceptual frame').
- $M_{pp}(T)$ is the set of all partial potential models of T, i.e. of those structures that satisfy the structural axioms referring only to the T-non-theoretical concepts of T (= the 'empirical basis').
- $r_T$ is the restriction function of T, i.e. a function assigning to each potential model its corresponding partial potential model by "cutting off" the T-theoretical concepts.
- $R_T$ is defined through $r_T$ but working at the next set-theoretical level: It assigns sets of partial potential models to sets of potential models.
- $K(T) = \langle M_p(T), M_{pp}(T), r_T, M(T)\rangle$ is called the 'core' of T. (We could omit mentioning $r_T$.)
- If K is a core, we obtain a 'specialization' K' of K (we write "K' σ K") if we add more proper axioms (laws) to K.
- I(T) is the 'domain of intended applications' (= the physical structures the theory is intended to deal with).
- A 'theory-element' (or a 'theory' for short) T is a pair $\langle K(T), I(T)\rangle$. To abbreviate, we will refer to all notions and structures belonging to K as 'the core level' and to notions related to I as 'the empirical level' of a given theory-element T.
- The 'empirical claim' of a theory $T = \langle K, I\rangle$ is a proposition of the sort: $I \subset R(M)$.
- If T is a theory-element, then T' is a 'specialization' of T (T' σ T) if K'(T) σ K(T) and $I'(T') \subset I(T)$.
- A structure consisting of a set of theory-elements partially ordered by the relation of specialization σ is called a 'theory-net'. What scientists call 'theories' are in most cases theory-nets in our sense. We advance the metatheoretical hypothesis that most, if not all, interesting cases of 'theories' in the intuitive sense, are 'tree-like' theory-nets, i.e. they have a _first_ theory-element of which all others are specializations. This first elements is called the 'basic theory-element'.

As for the more general symbolism used in this paper, we may remind of the following conventions.

- If T is a theory, then $x_T$, $y_T$, etc. are models of T. If x is a model, $t_x$ is a concept (set, relation, function) contained in the structure constituting x. Sometimes, when there is no danger of confusion, subindices will be omitted.

- If R is a relation, then $D_I(R)$ and $D_{II}(R)$ are its domain and range, respectively; $R^{-1}$ is its converse, and $R^2 = \{<x,y>/\exists z(<x,z> \in R \wedge <z,y> \in R)\}$.
- If f is a real function of time, then $D_t f$ is its derivative with respect to time.
- If S is a set, then $|S|$ is its cardinality, and $\Delta(S)$ its 'diagonal', i.e. the set of all pairs of identical elements of S.
- If q is a real number, then $|q|$ is its absolute value.

## I. APPROXIMATION THROUGH UNIFORMITIES

The basic notion we introduce to deal with approximation is that of an 'imprecision-set'[2]. Imprecision-sets in the present sense are elements of specific kinds of uniform structures, also called 'uniformities'. A set-theoretic axiomatization of the notion of uniform structure appears in Bourbaki (1951). Ludwig took it from there to apply it to the reconstruction of approximative relationships in physics. D. Mayr in his (1980) has followed him in his attempt to provide a general formalism to deal with approximation within Sneedian structuralism. Mayr's general intention is very similar to mine (though both arose independently). Nevertheless, Mayr's concrete development of his formalism is quite different from the present one. Perhaps both approaches will prove to lead to convergent results in the long run, but I still have not had the opportunity to investigate this question.

Normally, approximation is taken as a relation between single terms (mainly functions) of the same sort within a theory. Ludwig's and Mayr's approaches are no exception to this. However, this 'micrological' view[3] of approximation is too cumbersome and restrictive for the reconstruction of whole theories. It is more convenient to work 'globally' with whole models as units of comparison. It is not necessary to state what concrete terms of a theory are being approximated relationship. In our structuralistic approach it is more natural to work with models. We can define uniformities on classes of models. The "points" are then models. This has the additional advantage that even "qualitative" approximations can be formalized by means of the apparatus, as I argued with more detail in my previous articles on intratheoretic approximation (see Moulines, 1976a, and 1976b). Consequently, from our (informal) model-theoretic point of view, the most natural uniformity to work with is one defined on $M_p$ (= the class of potential models of a theory).

D1: $\mathcal{U}$ is a uniformity on $M_p$ iff

(1) $\emptyset \neq \mathcal{U} \subset \mathcal{P}(M_p \times M_p)$.

(2) For all $u_1, u_2$: if $u_1 \in \mathcal{U}$ and $u_1 \subset u_2$, then $u_2 \in \mathcal{U}$.

(3) For all $u_1, u_2 \in \mathcal{U}$: $u_1 \cap u_2 \in \mathcal{U}$.

(4) For all $u \in \mathcal{U}$: $\Delta(M_p) \subset u$.

(5) For all $u \in \mathcal{U}$: $u^{-1} \in \mathcal{U}$.
(6) For all $u_1 \in \mathcal{U}$, there is a $u_2$ such that $u_2^2 \subset u_1$ and $u_2 \in \mathcal{U}$.

The intuitive justification for each one of these axioms from the point of view of empirical approximation was already given in Moulines (1976a), pp. 211-212, and (1976b), pp. 46-47.

The elements of $\mathcal{U}$ we call 'imprecision-sets'.

Some useful properties that easily follow from the definition of a uniformity are these.

<u>T1:</u> $\emptyset \notin \mathcal{U}$.
<u>T2:</u> For all $u \in \mathcal{U}$: $u^n \subset u^{n+1}$.
<u>T3:</u> For all $u \in \mathcal{U}$: $u^n \in \mathcal{U}$.

Given our general view of the structure of physical theories, a natural question to ask at this point is whether a uniformity on $M_p$ induces a uniformity on $M_{pp}$ — that is, whether a "theoretical" approximation always produces a corresponding "non-theoretical" approximation as well. Intuitively, from the way we know empirical theories work, we would expect this to be the case. Actually, we can prove this only under one additional, though quite plausible, condition put on uniformities. The uniformities that satisfy this additional requirement we will call "empirical uniformities" — "empirical" in the sense that they appear to be meaningful only for empirical theories (that is, theories with a two-levels distinction of concepts). In order to define the notion of an empirical uniformity and to prove the "induction theorem" we first need some auxiliary concepts.

Let $\mathcal{U}$ be a uniformity and $u_i$ its imprecision-sets.

<u>D2:</u> $Rs(u_i) =: \{\langle y_1, y_2\rangle \,/\, \exists\, x_1, x_2 (y_1 = r(x_1) \wedge y_2 = r(x_2) \wedge \langle x_1, x_2 \in u_i)\}$.

<u>D3:</u> $\mathcal{V}[\mathcal{U}] =: \{v /\ \exists\, u \in \mathcal{U}: v = Rs(u)\}$.

Consider now those couples of potential models that have exactly the same T-non-theoretical components (in the strict sense that the non-theoretical functions take exactly the same values). They only differ in their T-theoretical part. Of these pairs of models, we could say, intuitively, that they are "empirically equivalent", since they contain or lead to the same empirical results. From the point of view of the empirical aspect of approximation, we could say that both models are quasi-identical, or also "as similar as we like". A mere change in the selection of the T-theoretical functions that does not have any implication for the T-non-theoretical determination of the physical system would not imply any change in the degree of approximation with which the system is described. To say it more precisely: For all $u \in \mathcal{U}$ and for any $x_1, x_2 \in M_p$, if $r(x_1) = r(x_2)$, then $\langle x_1, x_2\rangle \in u$. A uniformity satisfying this requirement we call an <u>empirical</u> <u>unifor-</u>

mity. In order to introduce this concept formally, it is convenient to define the concept of the "pseudo-diagonal" first.

D4: $\Psi(M_p) =: \{\langle x_1, x_2\rangle \ / \ x_1, x_2 \in M_p \wedge r(x_1) = r(x_2)\}$.

Consequently, an empirical uniformity will be a uniformity whose imprecision-sets always include the pseudo-diagonal.

D5: $\mathcal{U}$ is an empirical uniformity on $M_p$ iff
(1) $\mathcal{U}$ is a uniformity on $M_p$.
(2) For all $u \in \mathcal{U}$: $\Psi(M_p) \subset u$.

In an entirely analogous way we could define an empirical uniformity on $M_{pp}$. But it is obvious that a uniformity on $M_{pp}$ is always an empirical uniformity as well, since, in this case, $r(y) = y$.

We come now to the induction theorem.

T4: If $\mathcal{U}$ is an empirical uniformity on $M_p$, then $\mathcal{V}[\mathcal{U}]$ is a(n) (empirical) uniformity on $M_{pp}$.

Proof: We have to check each one of the six conditions for a uniformity.

(1) $\emptyset \neq \mathcal{V}[\mathcal{U}] \subset \mathcal{P}(M_{pp} \times M_{pp})$.
This easily follows from the premise and T1.

(2) $v_1 \in \mathcal{V}[\mathcal{U}] \wedge v_1 \subset v_2 \rightarrow v_2 \in \mathcal{V}[\mathcal{U}]$.
This is easy to prove by taking theoretical extensions $u_1, u_2 \in \mathcal{U}$ of $v_1, v_2$ with $u_1 \subset u_2$.

(3) $v_1 \in \mathcal{V}[\mathcal{U}] \wedge v_2 \in \mathcal{V}[\mathcal{U}] \rightarrow v_1 \cap v_2 \in \mathcal{V}[\mathcal{U}]$.
Take $u_1, u_2 \in \mathcal{U}$ with $v_1 = Rs(u_1)$, $v_2 = Rs(u_2)$.
From $\emptyset \neq u_1 \cap u_2$, it follows $\emptyset \neq Rs(u_1 \cap u_2)$. Further, it can be seen that $Rs(u_1 \cap u_2) \subset Rs(u_1) \cap Rs(u_2) = v_1 \cap v_2$,
Take $v_o =: Rs(u_1 \cap u_2)$. We have $\emptyset \neq v_o \subset v_1 \cap v_2$.
Since $u_1 \cap u_2 \in \mathcal{U}$, $v_o \in \mathcal{V}[\mathcal{U}]$ by definition and therefore also $v_1 \cap v_2 \in \mathcal{V}[\mathcal{U}]$ because of (2) above.

(4) $v \in \mathcal{V}[\mathcal{U}] \rightarrow \Delta(M_{pp}) \subset v$.
This trivially follows from the definitions.

(5) $v \in \mathcal{V}[\mathcal{U}] \rightarrow v^{-1} \in \mathcal{V}[\mathcal{U}]$.
This trivially follows from the definitions.

(6) For all $v_1 \in \mathcal{V}[\mathcal{U}]$ there is a $v_2$ with $v_2^2 \subset v_1$ and $v_2 \in \mathcal{V}[\mathcal{U}]$.
For any $v_1$ take a $u_1$ with $Rs(u_1) = v_1$.
By D1-(6) and T2 there is a t with $t \subset t^2 \subset u_1$.
If we apply D1-(6) to t again, there is a $u_2$ with $u_2 \subset u_2^2 \subset t$.
Therefore we get $u_2 \subset u_2^2 \subset t \subset t^2 \subset u_1$.

Let $v_2 = Rs(u_2)$. We want to prove that this $v_2$ is the one we are looking for. Take any two pairs connected by $v_2^2$:

$$\langle y_1, y_2 \rangle \in v_2, \quad \langle y_2, y_3 \rangle \in v_2.$$

Since $\langle y_1, y_3 \rangle \in v_2$, we want to show that $\langle y_1, y_3 \rangle \in v_1$.

Let $x_1, x_2, x_2', x_3$ be the T-theoretical extensions $y_1, y_2, y_3$, respectively, with

$$\langle x_1, x_2 \rangle \in u_2, \quad \langle x_2', x_3 \rangle \in u_2.$$

Since $\langle x_2, x_2' \rangle \in \Psi(M_p)$ and $\Psi(M_p) \subset u_2$, we obtain

$$\langle x_1, x_2' \rangle \in u_2^2 \subset t.$$

Since $u_2 \subset u_2^2$, $\langle x_2', x_3 \rangle \in u_2^2 \subset t$.

In sum, $\langle x_1, x_2' \rangle \in t$ and $\langle x_2', x_3 \rangle \in t$.

Therefore, $\langle x_1, x_3 \rangle \in t^2 \subset u_1$.

By applying the restriction operator Rs,

$$\langle y_1, y_3 \rangle \in v_1.$$

Imprecision-sets, which are in fact sets of pairs, i.e. relations, naturally induce corresponding "normal" sets, called neighborhoods, for each model. For any given $x \in M_p$, $u \in \mathcal{U}$, we define:

$$u(x) =: \{x' \in M_p \ / \ \langle x, x' \rangle \in u\}.$$

We call "u(x)" the "u-neighborhood" of x. Analogously, v-neighborhoods could be defined for any $y \in M_{pp}$. It is clear that $u(x) \subset M_p$ and $v(y) \subset M_{pp}$.

For a given x, there are many $u_i(x)$. Their collection we call U(x). And the union of all these U(x) we call U:

$$U =: \bigcup_{x \in M_p} U(x).$$

It is clear that $U(x) \subset P(M_p)$ and also $U \subset P(M_p)$. (In general, both inclusions will be _proper_ inclusions). U is what is frequently called a "base" for generating a standard topology on $M_p$.

We know by _T4_ that any empirical uniformity at the $M_p$-level induces an empirical uniformity at the $M_{pp}$-level. It can be easily seen that the same applies to the corresponding bases: A base U for a topology on $M_p$ induces a base V for a topology on $M_{pp}$.

In _de facto_ approximations in physical theories, not all imprecision -sets within a given uniformity $\mathcal{U}$ are used for purposes of application of the theory. Only a certain subset of $\mathcal{U}$ is allowed for actual use — leaving aside either "too large" or "too small", or other kinds of imprecision-sets that are inadmissible for some reason. In Moulines (1976a) the notion of an "admissible imprecision-set" within a uniformity $\mathcal{U}$ was introduced. The set of all admissible imprecision-sets for a given theory, call it $\mathcal{A}$, is a proper subset of $\mathcal{U}$. It should be clear that such a notion has a strong pragmatic component (referring to such things as conditions of experimentation and/or observation, size of the system studied, technological needs, availability of data, etc.) and is historically relative. Therefore, a formal definition of the admissible imprecision-sets could hardly be expected. However, some plausible necessary and formalizable conditions for it were set in that paper and it was also proven that the necessary conditions for admissibility of imprecision-sets at the $M_p$-level induce the corresponding conditions for admissibility at the $M_{pp}$-level.

Through the introduction of admissible imprecision-sets, one could develop a sort of "logic of approximation" by "blurring" the application of predicates and relations to structures. In particular, we can blur properties of models and relations between models once we have an appropriate uniformity $\mathcal{U}$ defined on them. To indicate this, we use a sort of model-theoretic operator $\sim$ and get the following definitions:

If x and y are two potential or partial potential models, P is a monadic predicate, and R a relation, then:

"$x \sim y$" means: There is an admissible inaccuracy-set u with

$$\langle x,y\rangle \in u.$$

"$P(\tilde{x})$" means: $\exists\, y(x \sim y \wedge P(y))$.

"$\langle x,y\rangle \in R$" means: $\exists\, x'(x \sim x' \wedge \langle x',y\rangle \in R)$,

and so on.

If X and Y are classes of potential models (or of partial models), then

"$X \sim Y$" means: $\forall\, x \in X\, \exists\, y \in Y(x \sim y) \wedge \forall\, y \in Y\, \exists\, x \in X(\ \sim y)$.

"$\overline{P}(\tilde{X})$" and "$\langle \tilde{X},Y\rangle \in \overline{R}$", etc. could be defined as before. (For more details, see Moulines, 1976a, p. 224.)

From a strictly formal point of view, this notation is not completely satisfactory: One should write the operator $\sim$ not on the arguments, but on the places, since one is blurring the predicates or relations, not the arguments themselves. However, this would give rise to a very cumbersome symbolism. The present notation, though not completely correct, does not imply any danger of confusion and is easier to handle.

## II. A GENERAL CONCEPT OF INTERTHEORETIC APPROXIMATION

We conceive of an intertheoretic relation as a relation between two theory-elements with different conceptual frames, i.e. with different sets $M_p$. And we know that a theory-element is defined by a core K (the "theoretical" part) and a domain of applications

I (the "empirical" part). Now, since the use of approximations is indispensable to apply any physical theory, I suggested in Moulines (1976a) that empirical uniformities should be considered as an essential component of any theory-element as well. Therefore, we propose to take as unit $T = \langle K, \mathcal{U}, I \rangle$, instead of $T = \langle K, I \rangle$, where $\mathcal{U}$ is an empirical uniformity defined on $M_p$ of K.

Any intertheoretic relation will be a relation between two such triples $\langle K, \mathcal{U}, I \rangle$ and $\langle K', \mathcal{U}', I' \rangle$. In the case of an <u>exact</u> intertheoretic relation, this will be a relation that involves only the sets $M_p$, $M_p'$ of the cores K, K'; that is, a relation of the form: $\rho \subset M_p \times M_p'$. Such a $\rho$ always induces (through the restriction function) an exact intertheoretic relation at the non-theoretical level:

$$\nu_\rho \subset M_{pp} \times M'_{pp}.$$

Now, the relations that bother us here are blurred or approximative intertheoretic relations (for short: "b.i.r."). We think of a b.i.r. as a relation between two different triples T and T' that not only involves the sets $M_p$ but also the uniformities $\mathcal{U}$, or to be more precise, the respective bases U and U'.

<u>D6:</u> For any two theory-elements T and T', a relation $\rho$ will be called a b.i.r. iff: $\rho \subset (M_p \times U) \times (M_p' \times U')$ with the requirement that, for any $x_1, x_2 \in M_p$, if $\langle x_1, u(x_2) \rangle \in D_I(\rho)$, then $u(x_1) = u(x_2)$, and for any $x_1', x_2' \in M_p'$, if $\langle x_1', u'(x_2') \rangle \in D_{II}(\rho)$, then $u'(x_2') = u'(x_1')$.

This last requirement intuitively means that a model x enters the relation only when accompanied of a "pertinent" neighborhood, that is, a neighborhood <u>of his</u>. I do not see any sense in allowing for a b.i.r. not satisfying this requirement — at least not in the present context.

It should be clear that any b.i.r. $\rho$ at the T-theoretical level induces a b.i.r. $\nu_\rho$ at the T-non-theoretical level: $\nu_\rho \subset (M_{pp} \times V) \times (M'_{pp} \times V)$.

Let us explain the intuitive idea behind this explication of a blurred intertheoretical relation. The intention is, roughly, that such a relation assigns at least one model x' of T' with its own neighborhood (its own degree of inaccuracy or its own blurring), u'(x'), to any model x of T with its own neighborhood, u(x). This can be understood as establishing the relation $\rho$ between the models of both theories through a blurring of the models on both sides of the relationship. Thus, in general, we will assume that, in order to establish a relation between two theories, a blurring on both sides of the relationship is needed. If one changes the degree of inaccuracy (the neighborhood u(x)) on one side, this will have some effect on the degree u'(x') on the other side — with the possible result that the intertheoretic relation does not hold any more.

However, the effect just described need not always occur.

It could be the case that no change in the degree of inaccuracy on one side (once the particular model x is fixed) modifies the degree of inaccuracy on the other side. For example, we assign a particular u'(x') to x through ρ independently of the u(x) we happen to use at the moment on the left side. Intuitively, this would mean that, in order to establish the b.i.r. ρ between T and T', only one side has to be blurred (in this example: the right side) — the blurring on the other side being only an "internal affair" of the theory T having no intertheoretic relevance. We are interested in introducing these possibilities formally. The way to do this is to use the idea that, if a neighborhood, say, u(x), is not relevant for ρ, we could put any other $u_i(x)$ in its place, and the b.i.r. would still hold.

<u>D7:</u> A b.i.r. ρ is at most a <u>right blurring</u> iff: For all $x \in M_p$, $x' \in M'_p$, $u_1(x), u_2(x) \in U$, $u'(x') \in U'$: if $\langle x, u_1(x)\rangle \ \rho \ \langle x', u'(x')\rangle$, then $\langle x, u_2(x)\rangle \ \rho \ \langle x', u'(x')\rangle$.

<u>D8:</u> A b.i.r. ρ is at most a <u>left blurring</u> iff: For all $x \in M_p$, $x' \in M'_p$, $u(x) \in U$, $u'_1(x') \in U'$, $u'_2(x') \in U'$: if $\langle x, u(x)\rangle \ \rho \ \langle x', u'_1(x')\rangle$, then $\langle x, u(x)\rangle \ \rho \ \langle x', u'_2(x')\rangle$.

These definitions of a right and a left blurring cover all potential models of both theories. But, if needed, we would introduce <u>restricted</u> right and left blurrings, which apply only to some subset $X \subset M_p$, or else right and left blurrings that only affect the partial models, that is, they are defined only with respect to $\nu_\rho$.

To abbreviate, when ρ is a right blurring we write "x ρ u'(x')" instead of "$\langle x, u(x)\rangle \ \rho \ \langle x', u'(x')\rangle$", to indicate that x is related to a specific blurring of x' namely u'(x'), independently of any blurring of x. Similarly, we write "u(x) ρ x'" for a left blurring. To abbreviate still more, when there is no need to specify the neighborhood of x that is involved in the blurring of ρ, we simply write "$x \ \rho \ \tilde{x}'$" for a right blurring and "$\tilde{x} \ \rho \ x'$" for a left blurring. If the blurring is neither left nor right, but both sides are involved, we write "$\tilde{x} \ \rho \ \tilde{x}'$".

We know that an exact intertheoretic relation ρ on $M_p \times M'_p$ induces an exact intertheoretic relation $\bar{\rho}$ on $\mathcal{P}(M_p) \times \mathcal{P}(M'_p)$. The same goes for blurred relations. A b.i.r. on $(M_p \times U) \times (M'_p \times U')$ induces a b.i.r. $\bar{\rho}$ on $(\mathcal{P}(M_p) \times \mathcal{P}(U)) \times (\mathcal{P}(M'_p) \times \mathcal{P}(U'))$. We will write, e.g. $\langle X, U(X)\rangle \ \bar{\rho} \ \langle X', U'(X')\rangle$, for any $X \subset M_p$, where $U(X) = \{u_i(x) / \ x \in X \wedge u_i(x) \in U(x)\}$, and analogously for U'(X'). Note that nothing precludes the possibility of having different neighborhoods $u_i(x)$, $u_j(x)$ for the same x in U(X).

In the same way we have defined right and left blurring for ρ, we could define them for $\bar{\rho}$ and write e.g. $X \ \bar{\rho} \ U'(X')$, and also $X \ \bar{\rho} \ \tilde{X}'$.

In the general case of a blurred intertheoretic relation we

have considered so far, a relation of this sort holds for any particular model x, respectively x', depending on the specific (admissible) neighborhood u(x), respectively u'(x'), we use to undertake a blurring of the model. However, in some physical theories, we can find more special cases of b.i.r.'s which, so to speak, "tend to exactness" — they hold not only for a particular neighborhood, but for any admissible neighborhood of x (or x'). In a way of speaking, we could say that in such cases, though the relation still is strictly speaking approximative, it "approaches" exactness with any admissible degree of inaccuracy — in other words, it is almost exact. Of course, this can occur on both sides of the relation.

D9: A b.i.r. $\rho$ is almost exact to the right iff: For all $x \in M_p$, $x' \in M'_p$, $u(x) \in U$, $u'_1(x') \in U'$, if $\langle x,u(x)\rangle \ \rho \ \langle x',u'_1(x')\rangle$, then, for any admissible $u'(x')$: $\langle x,u(x)\rangle \ \rho \ \langle x',u'(x')\rangle$.

For such a relationship we write "$x \ \rho \ \tilde{\tilde{x}}'$".

D10: A b.i.r. $\rho$ is almost exact to the left ("$\tilde{\tilde{x}} \ \rho \ x'$") iff: For all $x \in M_p$, $x' \in M'_p$, $u_1(x) \in U$, $u'(x') \in U'$, if $\langle x,u_1(x)\rangle \ \rho \ \langle x',u'(x')\rangle$, then for any admissible $u(x)$: $\langle x,u(x)\rangle \ \rho \ \langle x',u'(x')\rangle$.

We will say that a b.i.r. is almost exact on both sides if it is almost exact to the left and to the right. We write "$\tilde{\tilde{x}} \ \rho \ \tilde{\tilde{x}}'$".

As usual, these concepts could be relativized to specific subsets of $M_p$ and of $M_{pp}$. Also, they naturally induce the corresponding relations at the next set-theoretic level, e.g. almost exact right blurrings for classes of models. We would write e.g. "$X \ \bar{\rho} \ \tilde{\tilde{X}}'$" and "$\tilde{\tilde{X}} \ \bar{\rho} \ X'$".

The following corollaries express expected connections and are easy to prove.

T5: If a b.i.r. is a left blurring, then it is (at least) almost exact to the right — but not conversely.

T6: If a b.i.r. is a right blurring, then it is (at least) almost exact to the left — but not conversely.

Given the different sorts of structures that constitute a theory, the different sorts of intertheoretic blurring introduced above could combine in a number of ways to constitute a single intertheoretic relation. For example, we could have a $\rho$ that is a left blurring at the core level, but a right blurring for the domains of the intended applications; or else the left blurring could be undertaken at both the core and the applications levels; or a left blurring for the cores and an almost exact left blurring for the applications, and so on.

To deal more systematically with these different possibilities, for a b.i.r.:

| | left (1) | right (2) |
|---|---|---|
| K | $\rho^1$ | $\rho^2$ |
| I | $\rho_1$ | $\rho_2$ |

These possibilities can combine. For example, $\rho_2^1$ is a b.i.r. that is a left blurring at the core level and a right blurring at the applications level. By adding the qualification "almost exact" to each one of the former cases we obtain further possibilities. One of these cases will be relevant for the concrete case will discuss below: The Kepler-Newton relationship will prove to be a case of an almost exact b.i.r. $\rho_2^1$ of the reductive sort.

## III. APPROXIMATIVE REDUCTION

Leaving aside for the moment the whole problem-context of approximation and taking intertheoretic relations as if they would always be exact, we have every reason to assume that there are quite a number of different sorts of intertheoretic relations in physics. One of them is reduction. The paradigmatic example for this is the relation between rigid body mechanics and classical particle mechanics. (This example was formally reconstructed by Adams and Sneed.) Now, "reduction" is probably also a generic term covering a certain number of different subtypes of relations we could call of the reductive type. In this essay, we will consider one particular type of reductive relation and then go on to blur it.

If we forget about some complications which are irrelevant in the present context, the basic ideas underlying the general notion of reduction due to Adams and Sneed are the following.

A reduction relation $\rho$ between T and T' must be such that:

(1) $\rho \subset M_p \times M'_p$ with $M_p \neq M'_p$.

(2) $\rho^{-1}$ is a many-one function.

(Let us call a relation satisfying (1) and (2) a "quasi-reduction".)

(3) The laws of the reduced theory T can be derived from the laws of the reducing theory T'. Within a model-theoretic frame this can be expressed as:

if $\langle x,x'\rangle \in \rho$ and $x' \in M'$, then $x \in M$.

In the case $\rho$ would be defined over the whole of M' (a case that, however, does not seem to be the usual one in actual examples of reduction), we could also write this condition in the form

$\bar{\rho}^{-1}(M') \subset M$,

where $\bar{\rho}$ is the relation induced by ρ over classes of models. (This possibility will become important below.)

(4) The domains of applications I and I' are also in a ρ-correspondence, or, to speak more precisely, in the non-theoretical counterpart of the ρ-correspondence, which we call $\nu_\rho$.

I am not going to dwell upon the rationale for the basic ideas about structural reduction. The interested reader will find it in Sneed (1971) and Stegmüller (1973). A refined concept of reduction starting with the foregoing intuitions and taking account of all the complexities of the structuralistic approach appears in Balzer and Sneed (1977). For our present purposes, we need not take all these complications into account. The only result we need now is that, according to the standard explication of Balzer-Sneed, a reduction at the 'theoretical level' of $M_p \times M'_p$ always induces a reduction at the 'non-theoretical level' of $M_{pp} \times M'_{pp}$. If ρ is the reduction relation at the theoretical level, we will call $\nu_\rho$ the corresponding reduction relation at the non-theoretical level.

Let us also remind that a reduction relation ρ defined over single models induces a reduction relation $\bar{\rho}$ over classes of models. In the later case, we will write e.g. $\langle X,X'\rangle \in \bar{\rho}$ or $\langle Y,Y'\rangle \in \bar{\nu}_\rho$.

There may be different sorts of intertheoretic relations that in their own way fulfill the four requirements for a reduction. The particular conditions each one of these subtypes satisfy may differ more or less, may be stronger or weaker, more or less complicated, etc. But, in any case, they must all reflect those four requirements in some way. We are going to introduce a particular reductive relation now, and since we are intersted in approximation we will blur it also in a particular way that is convenient for our concrete purposes.

<u>D11:</u> If $T = \langle K,\mathcal{U},I\rangle$ and $T' = \langle K',\mathcal{U}',I'\rangle$ are two theory-elements, then $\langle T,\rho,T',I'_o\rangle$ is a $\rho\frac{1}{2}$ reductive approximation of T to T' iff:

(1) ρ is a quasi-reduction of T to T'.

(2) $I'_o \subset I'$.

(3) For any specialization H of K with $I \subset R(M(H))$, there is a specialization $H'_o$ of K' with $I'_o \subset R(M'(H'_o))$ such that $\bar{\rho}^{-1}(M'(H'_o)) \subset M(H)$.

(4) $\langle I,\tilde{I}'_o\rangle \in \nu_\rho$.

The intuitions behind this definition are as follows. T is the 'less developed' or 'poorer' theory that has to be reduced and T' is the 'more developed' or 'richer' theory that plays the reducing role. T is supposed to approximate T', but only within a proper subdomain $I'_o$, of the whole domain of applications I' of T'. In more intuitive terms, T' 'covers more' than T. $I'_o$ is the domain 'explained' by both T and T', but it is precisely only a subdomain of I'. This $I'_o$ stands in a ρ-correspondence with the <u>whole</u> domain

I of T, but not in an exact reductive correspondence: only in an approximate correspondence of the $\rho_2^1$ kind, as expressed in condition (4). Condition (3) means that all successful specializations of T have their approximate mirror image in T' in such a way that the axioms of each T-specialization follow from those of the corresponding T'-specialization in the model-theoretic sense suggested before.

We could obtain a concept of an almost exact $\rho_2^1$ reductive approximation instead of a (simple) $\rho_2^1$ reductive approximation by just substituting the operator $\approx$ for $\sim$ in the previous definition. Everything else would remain unchanged.

## IV THE RELATION BETWEEN KEPLER'S THEORY AND NEWTON'S GRAVITATIONAL THEORY-ELEMENT

In his (1973), Scheibe gave the particular conditions determining the relationship between Kepler's planetary theory and Newton's theory of gravitation. His analysis of this particular case proceeded in a totally concrete way, that is, he did not try to derive the conditions he found from a general intertheoretical scheme. The last section of the present essay offers a derivation of the particular conditions of the Kepler-Newton relationship from the general concept of a $\rho_2^1$-approximation. But before that, it is necessary for us to restate Scheibe's conditions in structuralistic terms in order to be able to include them in the present framework. This is what we proceed to do now.

The first step to be made is to identify the structures to be compared: those corresponding to Kepler's theory and those corresponding to Newton's. Contrary to some textbook opinions, we maintain that here we have a case of two really distinct theories, in the strong sense that not only the empirical laws contained in them but also the conceptual structures are different. They are 'theoretically incommensurable' in the sense of Stegmüller (1979). § 11 (see above). The non-theoretic structures of Kepler's and Newton's theory are of the same kind: kinematical descriptions of particle systems. But the theoretical 'superstructures' are different. To see this clearly, let us define both theories basic set-theoretic predicates determining the two corresponding sets of models[4].

D12: x is a Kepler system iff: There exist $\mathcal{P},\mathcal{T},s,\mu$ such that:

(1) $x = \langle \mathcal{P},\mathcal{T},s,\mu \rangle$.

(2) $\mathcal{P}$ is a finite, non-empty set.

(3) $\mathcal{T}$ is an interval of real numbers.

(4) s is a function from $\mathcal{P} \times \mathcal{T}$ into $\mathbb{R}^3$ such that s is twice differentiable with respect to the second argument.

(5) $\mu$ is a function from $\mathcal{P}$ into $\mathbb{R}$.

(6) There is a $p_o \in \mathcal{P}$ such that, for all $t \in \mathcal{T}$:

(a) $D_1^2 s(p_o,t) = 0 \wedge \mu(p_o) \neq 0$.
(b) For all $p \neq p_o$
  ($b_1$) $\mu(p) = 0$.
  ($b_2$) $D_1^2 s(p,t) = \mu(p_o) \cdot (s(p,t)-s(p_o,t)) \cdot |s(p,t)-s(p_o,t)|^{-3}$.
  ($b_3$) $1/2 \cdot |D_1 s(p,t)-D_1 s(p_o,t)|^2 - \mu(p_o) \cdot |s(p,t) - s(p_o,t)|^{-1} < 0$.

P is intended to be a set of particles (celestial bodies). T a time interval, s the position function, and μ the so-called 'Kepler-constant'. Condition (6) expresses an up-to-day formulation of 'Kepler's laws', where $p_o$ is intended to be the 'sun'[5]. The present formulation is taken from Scheibe (1973), p. 104, with slight modifications.

The set of all structures <P,T,s,μ> satisfying D12 is the set of models of Kepler's theory, M(Kep). We shall call it just K. Its set of potential models. $M_p$(Kep), is the set of all structures satisfying (1) to (5). The only Kep-theoretic concept here is μ. Therefore, the set of partial potential models $M_{pp}$(Kep) is the set of all structures satisfying (1) to (4).

D13: x is a Newtonian gravitational system iff: There exist P,T,s,m,f. such that:
(1) x = <P,T,s,m,f>.
(2) D3-(2).
(3) D3-(3).
(4) D3-(4).
(5) m is a function from P into $\mathbb{R}_+$.
(6) f is a function from P × T × N into $\mathbb{R}^3$ such that $\Sigma_i f(p,t,i)$ is absolutely convergent for every p,t.
(7) For all p ∈ P, t ∈ T: $\Sigma_i f(p,t,i) = m(p) \cdot D_1^2 s(p,t)$.
(8) There is a G ∈ $\mathbb{R}$ such that: for all p ∈ P, t ∈ T:

$$f(p,t,i) = -G \cdot \sum_{\substack{q \in P \\ q \neq p}} \frac{m(p) \cdot m(q)}{|s(p,t)-s(q,t)|^3} (s(p,t)-s(q,t))$$

The present formulation of a Newtonian gravitational system is a slight modification of the one to be found in Sneed (1971). Ch. VI. (Condition (7) expresses the 'Second Law' and (8) the 'Law of Gravitation'). M(New) is the set of structures satisfying D13. We shall call it just N. $M_p$(New) is the set of structures satisfying (1) to (6). As we know from Sneed's analysis, m and f have to be considered New-theoretical. Therefore, $M_{pp}$(New) is the set of structures satisfying (1) to (4). Since this set is the same as $M_{pp}$(Kep) let us take a common name for it: let us choose

'Kin' (= the set of kinematics). Then, the core of Kepler's theory will be Kep =:<$M_p$(Kep),Kin,K> , and the core of Newton's gravitational theory will be New := <$M_p$(New). Kin, N>.

Let us call $I_K$ the set of intended applications of Kep and $I_N$ the set of intended applications of New. We have $I_K \subset$ Kin and $I_K \subset$ Kin. Following Scheibe, we shall assume that $I_K$ contains only two-particles kinematics: this simplification also seems to be historically justified. Let us state this assumption explicitly in

P1: For all $y \in I_K$: $|P_y| = 2$.

This assumption does not imply any essential limitation of scope in our discussion. A further postulate we shall assume is this:

P2: $I_N \subset R(N)$.

That is, all intended applications of the Newtonian gravitational theory can actually be subsumed under the Newtonian core (they all satisfy the Second Law and the Law of Gravitation). Given the present state of our knowledge this assumption seems to be quite safe.

Notice that the whole of Kepler's theory consits just of the theory-element $T_K$ = <Kep,$I_K$>; it has no further specializations. The theory-net of Kepler's theory shrinks to a 'degenerated' net of just one element. On the other hand, the net of classical mechanics is certainly a huge theory-net. But we are only interested in one element of it, namely $T_N$ = <New,$I_N$>. Therefore, the intertheoretic comparison will be made between the single theory-element constituting Kepler's theory and one theory-element of the net of classical mechanics. We shall not have to bother about specializations.

As for the uniformities constituting both theories, at this point we will <u>not</u> determine them completely. We will not worry about the "internal affairs" of approximation within Newton's or Kepler's theory, nor about the content of the uniformities at the T-theoretical level. All we need to know is the form of the uniformity at the T-non-theoretical level for intertheoretic purposes. This is a uniformity $\mathcal{V}$ defined on $\mathcal{P}$(<u>Kin</u> × <u>Kin</u>) by the following condition.

Each member $v_\varepsilon$ of $\mathcal{V}_\varepsilon$ is defined for a given $\varepsilon \in \mathbb{R}_+$ as follows:

$\langle y,y'\rangle \in v_\varepsilon \in \mathcal{V}$ iff: For all $p \in P_y \cap P_{y'}$ and for all

$t \in T_y \cap T_{y'}$: $/s_y(p,t) - s_{y'}(p,t)/<\varepsilon$.[6)]

Whatever the uniformity at the T-theoretical level we choose (concerning masses and forces), it should be chosen in such a way as to induce this uniformity at the T-non-theoretical level.

According to the previous determination, we can define

$(I_N)_\varepsilon =: \{y / \exists y'(y' \in I_N \wedge \langle y,y'\rangle \in v_\varepsilon)\}$, and

$(I_K)_\varepsilon =: \{y / \exists y'(y' \in I_K \wedge \langle y,y'\rangle \in v_\varepsilon)\}$.

In a similar way, we could define $(R(N))_\varepsilon$ and $(R(K))_\varepsilon$.

As for the admissible imprecision-sets in both theories (independently of any further restrictions that may be put for "internal" reasons), in the present context we will admit at the T-non-theoretical level any $v_\varepsilon$ with $\varepsilon \neq 0$.

Let us come now to the reconstruction of Scheibe's analysis of the relationship between Kepler's laws and Newton's theory of gravitation. According to the conventions we have just introduced, Scheibe has arrived at the following results[7)].

(S 1) For all $\varepsilon > 0$: $I_K \subset (I_N)_\varepsilon$.

(This corresponds to Scheibe's condition (20) in Scheibe (1973), p. 114.)

(S 2) There is a specialization $H_i$ of N such that for all $\varepsilon > 0$: $R(H_i) \subset (R(K))_\varepsilon$.

(This corresponds to Scheibe's conditions (21), (22), p. 115).

(S 3) There is an $\varepsilon > 0$ such that: $I_N - (I_K)_\varepsilon \neq \emptyset$.

The last condition corresponds to no formally established condition in Scheibe's essay. Nevertheless, it corresponds to the intention of Scheibe's informal remarks on the 'superiority' of Newton's theory over Kepler's. There he says:

> *Für die approximativen Erklärungen [= reductive approximations in our sense] haben wir demgegenüber die Asymmetrie, daß die überwältigende Mehrheit von ε-Umgebungen von Newton-Fällen keinen Kepler-Fall enthält, während umgekehrt in jeder ε-Umgebung eines Kepler-Falles ein Newton-Fall zu finden ist (Scheibe, 1973, p. 116).*

It is to be noted, however, that the content of (S 3) is weaker than Scheibe's just quoted formulation, according to which 'the great majority' ('überwältigende Mehrheit') of Newtonian intended applications are not even ε-approximately Keplerian intended applications. Nevertheless, our weaker formulation in (S 3) is enough to guarantee the superiority of the Newtonian theory, and on the other hand it simplifies the discussion. We could presumably reconstruct Scheibe's stronger version by using a more sophisticated topological apparatus, as Scheibe himself has noted[8)], but I do not think that this complication would have any real bearing on the argument. It is enough to have the present weaker condition to show 'Newton's superiority' with respect to Kepler - and this is what we need.

## V. THE KEPLER-NEWTON RELATIONSHIP AS AN ALMOST EXACT REDUCTIVE APPROXIMATION

We have reached the point where ce can show the applicability of our general scheme of intertheoretic approximation to 'real-life' cases by deriving conditions (S1)-(S3) of the Kepler-Newton relationship from the general concept of a $\rho\frac{1}{2}$-approximation. The assumed intertheoretical assertion, from which (S1)-(S3) have to be deduced, is:

P3: $\langle T_K, \rho_{KN}, T_N, I_o^N \rangle$ is an almost exact $\rho\frac{1}{2}$ reductive approximation of $T_K$ to $T_N$ in $I_o^N$.

where $\rho_{KN}$ and $I_o^N$ are determined as follows.

(a) $\rho_{KN}$ is a quasi-reduction defined on $M_p(\mathrm{Kep}) \times M_p(\mathrm{New})$ satisfying condition (18a) given by Scheibe with respect to the theoretical magnitudes (Scheibe, 1973, p. 114):

$$m(p_o) \quad \left(1 + \frac{m(p_1)}{m(p_o)}\right)^{-2} = \mu(p_o),$$

and with respect to the non-theoretical concepts we are going simply to require identity under some specified circumstances. Hence, the induced non-theoretical quasi-reduction is:

$$\langle y_K, y_N \rangle \in \gamma_{\rho_{KN}} \quad \text{iff: } \exists\, x_K, x_K \in K \wedge x_N \in N \wedge y_K = r(x_K) \wedge$$
$$y_N = r(x_N) \wedge \langle x_K, x_N \rangle \in p_{KN} \wedge y_K = y_N.$$

It should be clear that $\rho_{KN}$ as well as $\gamma_{\rho_{KN}}$ satisfy the conditions for a quasi-reduction.

(b) $I_o^N$ is determined as $I_o^N = R(E_o^N)$, where $E_o^N$ is a specialization of N given by Scheibe's additional axioms as stated in his conditions (23), (24) of (1973), p. 115. Scheibe has shown that this $E_o^N$ is applicable to all cases of two-particles kinematics. The particular form of this specialization, however, is of no concern to us here. It suffices that it exists.

Now, we want to prove that (S1)-(S3) follow from the conditions of an almost exact $\rho\frac{1}{2}$ reductive approximation. First we need some lemmas which easily follow from the postulate and the definition of $\rho\frac{1}{2}$ approximation.

LEMMA 1: $I_o^N \subset I_N$.
This follows from P3 and <u>D11-(2)</u>.

LEMMA 2: $\exists\, H_i^N (H_i^N \subset N \wedge I_o^N \subset R(H_i^N) \wedge \overline{\rho}_{KN}^{-1}(H_i^N) \subset K)$.

This follows from P3, D11-(3), and the fact that the only specialization of K there is to take account of is K itself.

LEMMA 3: $\langle \tilde{\tilde{I}}_K, I_o^N \rangle \in \bar{\gamma}_{\rho_{KN}}$.

This follows from P3 and D11-(4).

T7: D11 and P3 imply (S1).

Proof: By Lemma 3 we have $\langle I_K, I_o^N \rangle \in \bar{\gamma}_{\rho_{KN}}$.

By definition of $\bar{\gamma}_{\rho_{KN}}$ this means

$$\forall y_k \in I_K \; \exists y_N \in I_o^N : \langle y_K, \tilde{\tilde{y}}_N \rangle \in \gamma_{\rho_{KN}}.$$

By definition of almost exact blurring, this implies:

$$\forall y_K \in I_K \; \exists y_N \in I_o^N \; \forall \varepsilon > 0 \; \exists y'_N \in I_N :$$
$$\langle y_K, y'_N \rangle \in \gamma_{\rho_{KN}} \wedge \langle y_N, y'_N \rangle \in V_\varepsilon .$$

Since $\gamma$ has been constructed as the identity, we get

$$\forall y_K \in I_K \; \exists y_N \in I_o^N \; \forall \varepsilon > 0 : \langle y_K, y_N \rangle \in V_\varepsilon .$$

By definition of $(I_N)$, it follows that

$$\forall y_K \in I_K \; \forall \varepsilon > 0 : y_K \in (I_N)_\epsilon , \text{ that is}$$

$$\forall \epsilon > 0 : I_K \subset (I_N)_\epsilon .$$

T8: D11 and P3 imply (S2).

Proof: By Lemma 2, there is an $H_o^N$ such that

$$H_o^N \subset N \wedge I_o^N \subset R(H_o^N) \wedge \bar{\rho}_{KN}^{-1}(H_o^N) \subset \tilde{\tilde{K}}.$$

The only member of this conjunction we really need here is the last one. From the already reported result of Balzer and Sneed (1977) we know that the 'theoretical' quasi-reduction $\bar{\rho}_{KN}^{-1}$ implies the 'non-theoretical' quasi-reduction $\bar{\gamma}_{pKN}^{-1}$ which 'mirrors' the correspondences stated by the first. In this case, this means

$$\bar{\gamma}_{pKN}^{-1} (R(H_o^N)) \subset R(\tilde{\tilde{K}}).$$

Since $\gamma_{pKN}$ is an identity, $R(H_o^N) \subset R(\tilde{\tilde{K}})$. By definition an almost exact blurring,

$$\forall y_N \in R(H_o^N) \; \exists y_K \in I_K \; \forall \varepsilon > 0 : \langle y_N, y_K \rangle \in V_\varepsilon .$$

By definition of $(R(K))_\varepsilon$,

$$\forall\, y_N \in R(H_o^N)\ \forall\, \varepsilon > 0:\ y_N \in (R(K))_\varepsilon.$$

This means:

$$\forall\, \varepsilon > 0:\ R(H_o^N) \subset (R(K))_\varepsilon.$$

From this we get (S2) by existential quantification over $H_o^N$.

T9: D11 and P3 imply (S3).

Proof: We shall assure the validity of (S3) if we can find a $z_N \in I_N$ such that $z_N \notin (I_K)_\varepsilon$ for some $\varepsilon$.

Let us choose a two-particles intended application of New of which we know that it is a physical part of a 'bigger' intended application of New: for example, the earth-moon system. (That such a system exists, of course, belongs to our background knowledge of $T_N$.) Let us call this kinematics $y_N$. Since the cardinality of $P_{y_N}$ is 2, we have $y_N \in I_o^N$. By hypothesis, we also know there is a $z_N \in I_N$ with $P_{z_N} \supsetneqq P_{y_N}$, therefore, the cardinality of $P_{z_N}$ is greater than 2. (For example, $z_N$ could be the earth-moon-sun system.)

Since $z_N \in I_N \subsetneqq R(N)$, we can apply the law of gravitation in particular to $z_N$. Now, it is a well-known exercise in mechanics to derive the paths of a two-particles system and the paths of the same particles when the system is extended to a three-particles system by means of the law of gravitation; we know that the paths of the particles in both cases will show non-negligible differences. Therefore, we obtain here:

$$(*)\quad \forall\, p \in P_{y_N}\ \forall\, t \in T_{z_N} \cap T_{y_N}:\ |s_{z_N}(p,t) - s_{y_N}(p,t)| \geq k,$$

for a given $k > 0$.

Now, let us take an $\varepsilon_o$ with $\varepsilon_o < k/2$. We choose $y_K$ with $y_K = y_N$, i.e. $y_K \in R(K)$. If $y_K \notin (I_K)_{\varepsilon_o}$, then the theorem is already proven by taking $y_N \in I_N$ instead of $z_N$. Suppose, however, it is the case that $y_K \in (I_K)_{\varepsilon_o}$. This means: $\exists\, y'_K \in I_K$: $\langle y_K, y'_K\rangle \in V_{\varepsilon_o}$. Suppose it would also be the case that $z_N \in (I_K)_{\varepsilon_o}$. Since $P_{z_N} \subsetneqq P_{z_N}$, it should also be the case, in particular, that $\langle z_N, y'_K\rangle \in V_{\varepsilon_o}$. Then, we have

$$\langle y_K, y'_K\rangle \in V_{\varepsilon_o} \wedge \langle z_N, y'_K\rangle \in V_{\varepsilon_o}.$$

This implies:

$$\forall p \in P_{y_K} \cap P_{y_K} \; \forall t \in T_{z_N} \cap T_{y_K}:$$

$$|s_{y_K}(p,t) - s_{y'_K}(p,t)| < \varepsilon_o \wedge |s_{z_N}(p,t) - s_{y'_K}(p,t)| < \varepsilon_o .$$

By elementary arithmetic,

$$|s_{y_K}(p,t) - s_{z_N}(p,t)| < 2_{\varepsilon_o} < k.$$

This last inequality contradicts the inequality in (*). Therefore, $z_N \notin (I_K)_\varepsilon$, while $z_N \in I_N$, as it had to be shown.

From the conjunction of these three theorems it follows that the Kepler-Newton relationship is a case of almost exact $\rho^1_2$ reductive approximation. Hence, this concept of intertheoretic approximation is adequate at least to this case.

NOTES

1) This article is a substantially revised and expanded version of my previous (1980). For the interested reader who already knows the previous version, I would like to remark that the two main innovations in the present article are the elimination of an 'intertheoretic uniformity' defined on $\underline{M}_p \cup \underline{M}'_p$ as an unnecessary complication and the development of a single schematic concept of intertheoretic approximation. I owe some helpful remarks to Professor Ignacio Jané, UAM (Mexico), and to many of the participants in the Osnabrück approximation colloquium concerning a first draft of this paper.
2) This term accords with the standard terminology adopted in this volume. In my (1980), I borrowed the term "inaccuracy-set" from D. Mayr in his (1980), where it is used in a similar context. Both are intended to be the English translation of Ludwig's "<u>Unschärfemenge</u>" (cf. Ludwig, 1978, § 6). In my (1976a), I used "fuzzy-set" instead of "inaccuracy-set". But I later noticed that this had led to some misunderstandings due to the somehow related but actually rather different use of the term "fuzzy-set" by L.A. Zadeh and his collaborators.
3) Following Stegmüller, I use the term "micrological" to refer to a general methodology which proceeds by term-by-term or by statement-by-statement comparison.
4) The reconstruction of Kepler's and Newton's theories offered here does not pretend to be historically accurate. The formulations given in this paper greatly differ from the original ones. But this point is of no real importance for our argument, not even from the diachronical point of view, since, within the structuralistic perspective, what matters is not the comparison

of statements (which can differ much through history), but the comparison of sets of models. And it is historically plausible to admit that the last have not changed much since Kepler's and Newton's times.

5) The equivalence between the equations in (6) and the classical formulation of Kepler's laws can be found in a textbook of mechanics.
6) Each imprecision-set $v_{\varepsilon}$ of the uniformity is determined by a particular $\varepsilon$.
7) The present formulation Scheibe's conditions is not literally identical to the formulation in the original text. For one thing, Scheibe's statements had to be translated into the jargon of our model-theoretic framework. Moreover, in his article Scheibe makes no explicit distinction between a T-theoretical and a T-non-theoretical level of concepts (though I think the distinction is somehow implicit in his treatment of the subject). However, these are minor differences of formulation, which affect the letter, not the spirit of the reconstruction. In a verbal communication to the author, Scheibe agreed on this 'translation' of his own terms.
8) In a verbal communication to the present author.

BIBLIOGRAPHY

Adams, E.W.: 1959, 'The Foundations of Rigid Body Mechanics and the Derivation of Its Laws from Those of Particle Mechanics", in The Axiomatic Method (ed. by L.Henkin, P. Suppes, A. Tarski). North-Holland, Amsterdam, 1959.

Balzer, W. and J.D. Sneed: 1977, 'Generalized Net Structures of Empirical Theories. Part I' Studia Logica 36 (1977), pp. 195-211.

Bourbaki, N.: 1953, Topologie générale, Paris, 1951.

Kuhn, T. S.: 1976, 'Theory-Change as Structure-Change: Comments on the Sneed Formalism', Erkenntnis, 10 (1976), pp. 179-200.

Ludwig, G.: 1978, Grundstrukturen einer physikalischen Theorie, Springer-Verlag, Berlin-Heidelberg, 1978.

Mayr, D.: 1980, 'Investigations of the Concept of Reduction, Part II', to appear.

Moulines, C.U.: 1976a, 'Approximate Application of Empirical Theories: A General Explication', Erkenntnis 10, (1976), pp. 201-227.

Moulines, C.U.: 1976b, 'Un concepto estructural de aproximación empîrica', Critica 24 (1976), pp. 25-51.

Moulines, C.U.: 1980, 'Intertheoretic Approximation: The Kepler-Newton Case', Synthese 45 (1980), pp. 387-412.

Scheibe, E.: 1973, 'Die Erklärung der Keplerschen Gesetze durch Newtons Gravitationsgesetz', in Einheit und Vielheit. Festschrift für Carl Friedrich von Weizsäcker (ed.

by E. Scheibe and G. Süssmann), Göttingen, 1973, pp. 98-118.

Scheibe, E.: 1976, 'Conditions of Progress and the Comparability of Theories', in Essays in Memory of Imre Lakatos (ed. by R.S. Cohen et al.). D. Reidel, Dordrecht, 1976, p. 547-568.

Scheibe, E.: 1979, 'Eine Fallstudie zur Grenzfallbeziehung in der Quantenmechanik', Unpublished manuscript, 1979.

Sneed, J.D.: 1971, The Logical Structure of Mathematical Physics, D. Reidel, Dordrecht, 1971, 2nd edition, 1979.

Stegmüller, W.: 1973, Theorienstrukturen und Theoriendynamik, Springer-Verlag, Berlin-Heidelberg, 1973, (translated into English as The Structure and Dynamics of Theories. Springer-Verlag, New York 1976).

Stegmüller, W.: 1979, The Structuralistic View of Theories, New York, 1979.

# SNEED'S THEORY CONCEPT AND VAGUENESS

W. Balzer

Seminar für Philosophie, Logik und Wissenschaftstheorie, Universität München
Ludwigstraße 31/I
8000 München 22
Federal Republic of Germany

## INTRODUCTION

The discussion of vagueness and approximation in empirical and especially physical theories - as far as it is intended as a part of meta-science - faces the following problem. There are practically no utterances of practising scientists about the relation of theory and approximation which could serve as 'data' against which the meta-discussion might be 'tested'. This is not to say that approximation is not relevant for physics: the very first laboratory courses in physics prove the contrary. The calculation of errors in measurement and the 'theory of disturbances' are essential for physical methodology. But these achievements are not suited to clarify the relation between theories and reality.

Consequently the meta-discussion cannot proceed 'empirically', i.e. by formulating hypotheses for which 'real life' examples are presented. Rather the discussion is 'analytic' in the sense that it analyses the connection between theory and reality and the role of approximation on the basis of some general background. As long as we have no case studies of concrete examples of approximation the discussion will remain analytic, and what I shall have to say in the following will be analytic to a large extent, also.

There are at least three proposals for treating vagueness or approximation technically. A first, 'classical' text here is Przelecki's [9]. He offers an apparatus in terms of the semantics of first-or higher-order theories. A second proposal is due to Ludwig who introduces topological uniformities to 'smear over' the sharp theoretical images of the theory (see hin contribution in this volume, or his [7]). Thirdly, Moulines in [8] has applied Ludwig's ideas to Sneed's theory concept.

My aim in this paper is to consider these propositions with respect to their underlying intuitions and to look for adequate formal realisations of these intuitions in Sneed's concept of an empirical theory.

It should be noted that the latter concept is 'open' enough to allow for an 'emendation' by building in aspects of approximation without essentially changing the original content. I think such 'openess' is favourable for meta-theoretical concepts and fits the idea propounded by Stegmüller in [14] of starting an analogue to the Bourbaki-programme in the realm of empirical sciences on the basis of some suitable theory concept. In order to build in approximation into Sneed's theory concept it seems necessary to start with a brief discussion of some central features of this concept.

## I. SNEED'S THEORY CONCEPT

I will concentrate on the aspect of formulating empirical claims which is especially relevant for problems of approximation. The discussion of empirical claims will force me to look at the so called 'problem of theoretical terms' in some detail. For a broader discussion the reader is referred to [12], [13], [14] and for an appropriate technical formulation especially to [3].

The starting point of Sneed's theory concept is that there are axiomatizations of physical theories. Any such axiomatization yields a class of structures or (in the following) models: M. In special contexts it is profitable to regard M as the class of structures of a certain species in the sense of Bourbaki [4] (as proposed by Ludwig [7]) or as the class of structures of a certain 'similarity type' in the sense of Feferman [5].

But the characterization of a class of models has nothing to do with experience or reality. We have to ask how M can be used in order to formulate statements about reality or empirical claims. It is clear that this cannot be done without reference to 'real systems'. So let us introduce a set I such that the members of I are 'real systems'. In classical mechanics, for instance, such real systems are the solar system and its sub-systems, concrete pendulums, the tides, and concrete harmonic oscillators. Such real systems will be called intended applications, so I is the set of intended applications. As soon as we have these two components, M and I, we can try to formulate an empirical claim of the following form: "All intended applications of the theory are models of the theory". This claim admittedly is a bit unclear, in particular it is not clear what is meant by 'are'. The meaning of 'are' depends on the structure of the intended applications. There are various possibilities of giving a special structure to the intended applications of which I want to point out two extremes. A first possibility is to regard an intended application as a (finite) set of atomic propositions[1]. (Proponents of this view are, e.g., Ludwig and Popper). A second possibility is to regard an intended application as a structure of the same 'type' as the members of M, i.e. intended applications are structures wich possibly might be models. A discussion and comparison of these possibilities which, I think, arise from different strategies in reconstructing a theory or a 'hierarchy' of theories is beyond the bounds of this paper. As I want to concentrate on Sneed's theory concept I shall follow Sneed and choose the second possibility.

A theory T then consists of at least three components, M, I and $M_p$, where $M_p$ is the class of potential models, i.e. of structures of the type of models (which need not satisfy the central axioms of the theory).

D1 a) T is a theory* iff $T = \langle M_p, M, I \rangle$ and

1) $M_p$ and M are classes of structures of the same type

2) $M \subseteq M_p$

3) $I \subset M_p$

b) the empirical claim* of $T = \langle M_p, M, I \rangle$ is that $I \subset M$

A claim of this form immediately leads to the problem of theoretical terms if we ask how the claim can be tested. Since the claim has the form of a universal sentence

$$\forall x (x \in I \Rightarrow x \in M)$$

it is sufficient to consider one arbitrary $x_o \in I$. How do we test

(1) $x_o \in I \Rightarrow x_o \in M$ ?

If we know that $x_o \in I$ we know that $x_o$ has the form of a potential model. Let us assume the potential models of T having the form $x = \langle y_1, \ldots, y_n \rangle$ where $y_1, \ldots, y_n$ are functions or relations. Then $x_o$ also has this form: $x_o = \langle y_1^o, \ldots, y_n^o \rangle$. But we do not know what the various $y_i^o$ look like. The set I is not defined explicitly but 'paradigmatically', i.e. some concrete systems are singled out - forming a finite set $I_o$ of 'paradigm intended applications' - and other systems are said to belong to I if they are 'sufficient-

1) I speak of propositions rather than of sentences in order to preserve the distinction between models and axioms, or between interpretation and language. Propositions (in the technical sense) are entities of the first kind. They correspond to sentences at the level of language, and by an atomic proposition the reader may understand a proposition corresponding to an atomic sentence.

ly similar' to members of $I_o$. In order to test (1) we must describe $y_1^o, \ldots, y_n^o$ and this in turn leads to measurements in order to find out the 'values' of $y_1^o, \ldots, y_n^o$.

Now we have the following problem. There are theories T with theoretical components. That is, there is some component, say $y_n$, in the potential models of T such that every measurement of $y_n$ presupposes some version of T's axioms to be satisfied in the course of the measurement (for details compare (2)).

As an example take mass in classical particle mechanics (CPM). The potential models of CPM have the form $\langle P,T,s,m,f\rangle$, where P is a non-empty, finite set (of 'particles'), $T \subset \mathbb{R}$ is an open interval (time), and s (position), f (force), m (mass) are such that $s : P \times T \to \mathbb{R}^3$ and $f : P \times T \times \mathbb{N} \to \mathbb{R}^3$ are smooth and $m : P \to \mathbb{R}_+$. Models of CPM are those potential models which satisfy Newton's second law:
$\forall p \in P \forall t \in T(m(p)\ddot{s}(p,t) = \sum_i f(p,t,i))$. Mass in this theory is a theoretical component. To prove this statement we have to consider all methods of measurement for mass. Since this totality is very difficult to describe we can only try to confirm it by means of some examples. Let us consider the measurement of mass by means of central, inelastic collisions. This method yields the mass ratio in terms of differences of velocities, provided the two particles move on a straight line. But these assumptions do not guarantee that we really measure mass. If, e.g., the whole measuring device is accelerated along the line of motion we will not accept the ratio of the velocity differences as an expression being equal to the mass ratio although the performance of the experiment still might be possible. A physicist will say that of course the measurement has to be performed in an inertial system. If by this he means what Prof. Ludwig has suggested, namely that objects made out of 'soft' material do not change when submitted actively to the usual Galilei transformations then he already presupposes 'part of' Newton's second law, namely that

(2) $\forall p \in P \forall t \in T(\sum_i f(p,t,i) = \varphi(m(p)\ddot{s}(p,t)))$ with

$\varphi : \mathbb{R}^3 \to \mathbb{R}^3$ and $\varphi(0) = 0$.

For if the objects are not accelerated the presence of non-zero forces would deform them. Among those functions $\varphi$ for which (2) is true identity certainly is the simplest one, and so we can say that some version of Newton 2 is satisfied.

The reader may check other methods for measuring mass and see whether he can find an example where no version of Newton's second law is necessary. But in all cases he must be able to convince us that he really measures the 'mass of CPM'. Part of this task will always be to exclude counterexamples of the kind just discussed, and usually such counterexamples lead to the assumption of some version of Newton's second law. Anyway, Sneed and others have claimed that mass is CPM-theoretical. This is an 'empirical claim' at the meta-level, and, I think, a correct one.

Usually the most interesting terms of empirical theories turn out to be theoretical. Thus the following 'problem of theoretical terms' cannot be neglected: if T has a theoretical component then confirmation of an empirical claim of the form of D1-b) leads to a version of circularity. For in order to check, say, whether $x_o \in I \Rightarrow x_o \in M$ is true we have to determine the components of $x_o$ among which there is a theoretical one. In order to determine (to measure) this component it is presupposed that the 'measuring-system' already satisfies a version of the axioms. Thus in order to determine whether $x_o$ is a model we already presuppose that some $x_1$ satisfies a version of T's axioms. So we only have a kind of conditional confirmation, namely relative to the assumption that $x_1$ is a 'model-version' of T. If we want to transform this conditional confirmation into an 'absolute' one we have to test whether $x_1$ in fact is such a version. But this leads to a repetition of the whole story. In order to test whether $x_1$ satisfies a version of T's axioms we first have to measure $x_1$'s components, and since $x_1$ contains a theoretical component this is possible only by presupposing some $x_2$ being a version of T's axioms.

I agree with the objection that this formulation of the problem of theoretical terms is still a bit vague: what is a 'version of T's axioms' and what is a 'measuring-system'? The latter notion has been clarified in [2] where it is called 't-determining model', while the concept of a version still waits for precise explication. But such an explication does not pose any principal question and I avoid it only because it leads to rather complicated technical formulations.

An elegant solution of the problem of theoretical terms was suggested by Ramsey: just reformulate the empirical claim such that the theoretical components come under existential quantifiers. For given $x_o$ then it is no longer claimed that $x_o$ is a model but that <u>there exist</u> theoretical components which, if adduced to $x_o$, yield a model. Certainly this solution (which is called the solution by Ramsey-sentences) is not the only possible one and it seems to be a rather coarse solution. But it fits with practising scientists' answers to questions like: "What does the theory tell us about this system?".

A more precise formulation forces us to extend our theory concept as follows. We have to introduce a distinction between theoretical components and non-theoretical components of the potential models. We can do this by making explicit the structure of potential models (a potential model x then being of the form $x = \langle y_1,\ldots,y_n \rangle$) and by introducing a natural number $m (m \leq n)$ together with the stipulation that $y_1,\ldots,y_m$ are non-theoretical components and $y_{m+1},\ldots,y_n$ are theoretical components of x. The class of structures consisting of non-theoretical components only is called $M_{pp}$ and its members 'partial potential models'. In this frame it is sufficient to require I to be a subset of $M_{pp}$ instead of $M_p$. This, finally, allows us to formulate an empirical claim in which no theoretical component occurs as constant (D2-b) below).

D2 a) T is a theory[1] iff $T = \langle M_p, M, M_{pp}, I\rangle$ and

1) M and $M_p$ are classes of structures x of the same type and of the form $x = \langle y_1, \ldots, y_n\rangle$

2) $M \subset M_p$

3) there is $m \leq n$ such that

$$M_{pp} = \{\langle y_1, \ldots, y_m\rangle / \exists y_{m+1} \ldots \exists y_n (\langle y_1, \ldots, y_m, y_{m+1}, \ldots, y_n\rangle \in M_p)\}$$

4) $I \subset M_{pp}$

b) the empirical claim[1] of $T = \langle M_p, M, M_{pp}, I\rangle$ is that

$$\forall x (x = \langle y_1, \ldots, y_m\rangle \in I \Rightarrow \exists y_{m+1} \ldots \exists y_n (\langle y_1, \ldots, y_m, y_{m+1}, \ldots y_n\rangle \in M))$$

This claim can be written more elegantly if we use the function $r : M_p \rightarrow M_{pp}$ defined by $r(\langle y_1, \ldots, y_m, y_{m+1}, \ldots, y_n\rangle) = \langle y_1, \ldots, y_m\rangle$. The claim then is that $\forall x (x \in I \rightarrow \exists x' (r(x') = x \wedge x' \in M))$ or, still more briefly

(3) $I \subset r(M)$.

This claim no longer creates a problem with theoretical terms for the role which these play in (3) can be tested by paper and pencil operations: we only have to check whether there exist theoretical augmentations x' such that $x' \in M$. The determination of x - which is treated as 'given' - is unproblematic, at least from the point of view of the theory under consideration, because x does not contain any theoretical component.

I do not treat here other features of Sneed's theory concept, as for instance the 'constraints' which are crucial to arrive at non-vacuous empirical claims, or the notion of a theory-net which allows us to treat dynamical aspects as well as larger 'portions of theory'.

## II. VAGUENESS AS MULTIPLICITY

This view of vagueness as well as the slogan is due to Przelecki. In [9], [10] and [11] he treats vagueness in a semantical frame. There is given a language L of an empirical theory which has to be interpreted (in the sense of formal logics). In contrast to formal or mathematical theories the interpretations of an empirical language cannot be arbitrary. Only certain 'intended' interpretations are of interest. Vagueness now means that there is not one unique intended interpretation for L but there are many 'admissible' intended interpretations.

The idea behind this approach is that an interpretation of language L is given or induced by ostensive or non-verbal means, in physics for instance by means of measurement. And it is a fact that two interpretations of some expression of L as given by measurements usually are different.

For example, if we consider the function m of CPM and interpret it, first, as the mass function of a given system x at time t - which is determined by means of mass-measurements - and, second, as the mass function of the same system x at time t + 1 again determined by means of measurement, then usually the two interpretations will be different. In other words, if we twice measure the mass of some particle p then, in general, we obtain two different values. This is an observation about how the world is and little or even no theory is needed to establish such observations.

Thus we have as a general feature of most terms of empirical theories that their interpretation at different occasions will be different. This gives rise to a kind of vagueness: the empirical claim will depend on the special way in which we arrive at an interpretation of the terms occuring in this claim.

To avoid misunderstandings it seems useful to add two remarks. First, this kind of vagueness has nothing to do with the question of how precise our measuring instruments are. If they are not very precise then of course we will expect different values at different times. But even if they are very precise, say, more precise than our ability to resolve visual impressions, the situation does not change. For in such cases by means of ingenious devices we can make 'visible' (e.g. countable) the differences which are too fine for the senses. And with such auxiliary devices again we obtain different results in different performances. Second, this kind of vagueness does depend on the kind of measurement which is used. If the term to be interpreted for instance is a function taking only the values 0 and 1 (think of yes-no experiments in quantum mechanics) then there are examples in which practically all performances of the experiment yield the same value and thus the same interpretation. But such experiments are important mainly in quantum physics and there we have statistical features elsewhere. In classical theories we could introduce such measurements but this would necessitate a substantial reformulation of the established theories. I do not want to speculate on what could be the case. It is the case that the terms of the established classical theories, if interpreted by means of measurement, yield multiple results. (The point here is that 'smeared' terms, lime $m \pm \varepsilon$, are not terms occurring in established theories.)

Now in order to build this kind of vagueness into the Sneedian theory concept we have to look for counterparts of a language L and of 'interpretations'. A language L is present only implicitly in a theory T in the sense of D2-a). The different components of the potential models correspond to the different basic symbols of a language. More precisely, we could regard these components as given by some interpretation of a language containing a non-logical constant for each such component. The intended interpretations mentioned above in T correspond to the set of intended applications I. For the other components, M, $M_p$ and $M_{pp}$ are classes of all structures of a certain kind. So they are, so to speak, classes of all possible interpretations of suitable languages (in case of M the class is further restricted by purely verbal or formal requirements). So I is the only candidate to express that we want to deal only with certain intended objects (interpretations or applications). But the correspondence of an intended interpretation with an intended application seems to be sound also on intuitive grounds. An intended interpretation in fact is some structure consisting of components like those of intended applications. The distinction between theoretical and non-theoretical components does not create any problems because Przelecki also has such a distinction and the intended interpretations at first are considered on the non-theoretical level.

If we identify intended interpretations with intended applications then a multiplicity of interpretations becomes a multiplicity of intended applications. This multiplicity must not be confused with a multiplicity of intended applications already present in D2-a), namely the multiplicity constituted by the fact that usually I contains more than one member. Different elements of I in D2-a) stand for different real systems, like two systems consisting of two different concrete pendula or another concrete system which is an harmonic oscillator. The multiplicity we are discussing here is a multiplicity of intended applications corresponding to one single real system. The idea is that corresponding to some given real system we always have a multiplicity of intended applications (intended interpretations) created by the fact that the determination of the components of the given system yields different results at different occasions.

These considerations give rise to the following formal treatment. We impose new structure on the set I of intended applications of D2-a). We distinguish a family of sets of intended applications $(I_j)_{j \in J}$ such that for all j: $I_j \subset M_{pp}$. The indices in J are thought of as representing concrete systems, i.e. they are names of concrete systems. For instance, the pendulum swinging at a certain time in a certain laboratory will be denoted by an index in J. The set $I_j$ corresponding to this index then contains the various interpretations which we obtain for this pendulum if we determine its position function and the mass of the particle at different occasions. I is given as the union of all these $I_j$.

It may be noted that a set $I_j$ does not contain a distinguished element, the 'true' description of the system. This is no shortcoming but an advantage. There simply are no distinguished descriptions of concrete systems. This point can be better understood by contrasting it with the opposite one. According to the opposite point of view for a concrete system, like the pendulum, given ostensively, there exists one distinguished, 'sharp' and 'true' theoretical image or description, and the set $I_j$ is obtained by smearing this distinguished structure which is necessary because of our limited ability to determine this true object. I think this latter view is wrong and misleading. It is wrong as concerns the everyday experience in measuring, and it is misleading in metatheoretic thinking because it strengthens our tendency to overestimate theoretical pictures and to underestimate their accessibility.

I now go beyond what Przelecki has proposed by asking whether we can (or should) introduce uniform structures at this stage. Of course we can. The sets $I_j$ are natural condidates on which we can impose uniform structures $\mathcal{U}_j$. The members U of such an $\mathcal{U}_j$ are subsets of $I_j \times I_j$. $\langle x,x'\rangle \in U$ with x, x' $\in I_j$ and $U \in \mathcal{U}_j$ means that x and x' are similar of 'degree U', and U will be called a degree of similarity in the following. There are two arguments in favour of such $\mathcal{U}_j$. The first argument is that in case of quantitative theories, which is the normal case in physics, we will have one or several natural mathematical uniformities on each $I_j$. A second argument is that in physical practice we find activities corresponding to such uniform structures. Calculations of errors, for instance, consist of calculations of mean-values and of variances, and this can be interpreted as calculating a degree of similarity U to which both systems - the one given by the mean-values and the other one given by the measured values - belong. We will therefore, in addition, introduce uniformities with each set $I_j$ (see D4) below.

If we introduce uniformities we somehow must restrict the possibilities for dissimilarity in order to avoid smearing the empirical claim to such an extent that it becomes logically true. The usual definition of uniform structures contains an axiom of the form $U \subset V \wedge U \in \mathcal{U} \Rightarrow V \in \mathcal{U}$ (compare D3-b) below for an understanding of this formula). Thus the uniformity has to contain with each U all V representing smaller degrees of similarity ($U \subseteq V$). This axiom implies that the cartesian product of the basic set always belongs to the uniformity, i.e. that the uniformity $\mathcal{U}$ contains a degree of similarity according to which every object of the basic set is similar to every other such object. Clearly we need some restriction here. There are two possibilities. First, we can drop the axiom just mentioned and work with 'weak' uniform structures as defined in D3-b) below. In these structures the above axiom need not be satisfied and we can think of the uniformity $\mathcal{U}$ itself as representing the 'admissible' degrees of similarity and no others. In this case we can regard the members of $I_j$ as being all admissible in the sense that they are sufficiently similar to each other. A second possibility (used by Moulines on the theoretical level) is to treat $I_j$ as full uniform space and to choose a subset $\mathfrak{A}_j$ of each $\mathfrak{U}_j$ with the idea that $\mathfrak{A}_j$ contains the admissible degrees of similarity. For reasons of simplicity I will take the first alternative.

We now can introduce the concept of a weak uniform space.

D3 a) if N is a set and $U \subseteq N \times N$ then

a.1) $\Delta_U = \{\langle y,y\rangle / \exists z \in N(\langle y,z\rangle \in U) \vee \exists z \in N(\langle z,y\rangle \in U)\}$

a.2) $U^{-1} = \{\langle y,z\rangle / \langle z,y\rangle \in U\}$

a.3) $U^2 = \{\langle x,y\rangle / \exists z(\langle x,z\rangle \in U \vee \langle z,y\rangle \in U)\}$

b) x is a <u>weak uniform space</u> iff $x = \langle N,\mathfrak{U}\rangle$ and

1) N is a non-empty set

2) $\emptyset \neq \mathfrak{U} \subset Pot(N \times N)$

3) $\forall U,U'(U\in\mathfrak{U} \wedge U'\in\mathfrak{U} \Rightarrow U\cap U'\in\mathfrak{U})$

4) $\forall U(U\in\mathfrak{U} \Rightarrow \Delta_U \subset U \wedge U^{-1}\in\mathfrak{U})$

5) $\forall U'\exists U(U'\in\mathfrak{U} \Rightarrow U^2\subset U' \wedge U\in\mathfrak{U})$

For an interpretation of these axioms see, e.g. Ludwig's contribution to this volume.

D4 T is a <u>theory</u> iff $T = \langle M_p, M, M_{pp}, I, (I_j)_{j\in J}, (\mathfrak{U}_j)_{j\in J}\rangle$ and

1) $\langle M_p, M, M_{pp}, I\rangle$ is a theory$^1$ (see D2-a)

2) J is a non-empty set

3) for all $j \in J$ : $I_j \subset I$ and $\mathfrak{U}_j \subset Pot(I_j \times I_j)$

4) $I = \bigcup_{j\in J} I_j$

5) for all $j \in J$ : $\langle I_j, \mathfrak{U}_j\rangle$ is a weak uniform space

As already said, J is a set of names of concrete physical systems and each $I_j$ is a set of intended interpretations arrived at by measuring system j. Each $\mathfrak{U}_j$ is a weak uniformity on $I_j$.[2)]

We now can formulate a vague empirical claim of T. To this end we agree on the following notation. If $X \in \prod_{j\in J} I_j$ ($\mathfrak{X} \in \prod_{j\in J} Pot(I_j)$) then $\rho(X) = \{X(j)/j\in J\}$ ($\bar{\rho}(\mathfrak{X}) = \bigcup\{\mathfrak{X}(j)/j\in J\}$). Intuiti-

---

2) We might add $I_i \cap I_j = \emptyset$ for $i \neq j$ but this seems a rather strong requirement.

vely, $\rho(X)$ is the set of 'components' $X(j)$ of X.

D5 the vague empirical claim[2] of $T = \langle M_p, M, M_{pp}, I, (I_j)_{j\in J}, (\mathcal{U}_j)_{j\in J}\rangle$ is that

$$\exists X (X \in \prod_{j\in J} I_j \land \rho(X) \subset r(M))$$

The vague empirical claim[2] formulated with the help of this apparatus intuitively says that there exists a combination of intended applications X such that for each name j, X contains exactly one member, and all these members by means of theoretical components can be extended to models. In other words, for each concrete system there exists an intended description (given by measurement) and these descriptions can be extended to models.

Some further remarks may be helpful. First, we observe that the uniformities $\mathcal{U}_j$ in a natural way induce a uniformity $\mathcal{U}$ on $\prod_{j\in J} I_j$. The present account then formally can be regarded as smearing an empirical claim[1]: $I^* \subset r(M)$ by smearing $I^*$ with the help of $\mathcal{U}$. Here, $I^*$ is the set of 'sharp' intended applications in the sense of D2). In the terminology of D5) we would say that for some 'ordering' of $I^*$, say $I_o(\rho(I_o)=I^*)$, $I_o \in \prod_{j\in J} I_j$. But as already said this point of view is misleading for it suggests the existence of some distinguished member of $\prod_{j\in J} I_j$, namely some such $I_o$.[3]

Second, we note that the claim[2] of D5) contains two 'Ramseyfications', two existential quantifications. For the part $\rho(X) \subset r(M)$ implicitly contains another existential quantifier. So the explicit formulation is this:

$$\exists X \exists Y (X \in \prod_{j\in J} I_j \land r(Y)=\rho(X) \land Y \subset M).$$

As a third remark let us pose the question why we have chosen an existential quantification for X. Alternatively we might have put X under an universal quantifier, thus requiring that all combinations of $\prod_{j\in J} I_j$ can be extended to models. But such a claim would be false in most cases. To see this consider for instance CPM enriched by the law of gravitation. For suitably given paths (expressed by s) we can find masses such that the gravitational equations are satisfied. But in an arbitrarily small neighbourhood of s we find mathematical paths s' such that there are no masses for which s' satisfies the gravitational equations. We cannot exclude - and in fact it is very likely - that both these paths s and s' occur in one $I_j$, i.e. they both arise from a concrete

---

3) This argument was put forward by Przelecki in a discussion.

system by means of measuring the corresponding paths. In this case a universally quantified claim is false. This seems reason enough to justify the existential quantifier.

Fourth, it is easy to see - and has been pointed out by H.-J. Schmidt - that the claim[2] of D5) does not at all refer to the uniformities $(\mathcal{U}_j)_{j\in J}$ occuring in T. In fact, as long as we choose exactly one member from each $I_j$ there is no possibility of using some $U_j \in \mathcal{U}_j$ in formulating the claim[2]. The component $(\mathcal{U}_j)_{j\in J}$ assigned to the theory (c.f. D4) thus seems redundant. But I think that the intuition of adding these uniformities is sound: there is something like degrees of similarity between different 'sets of data' obtained by measuring the same system several times, and these similarities are relevant and highly important for connecting theory with experiments. The inadequacy of D5) - as revealed by its neglecting the $\mathcal{U}_j$ - is this. We have oversimplified the claim by choosing exactly one member from each $I_j$. If the indices in J are to denote concrete physical systems then of course there is the possibility of measuring one such system j at several times and thus obtaining several members of $I_j$ to be regarded in the empirical claim. And only in this case do we have the opportunity of investigating those different members of $I_j$ with regard to their respective similarities. Such similarities, once established quantitatively, form a kind of standard with which new measurements or measurements of similar systems are compared. Such procedures are the basic - and, I think, even the only - means to obtain and explicate something like confirmation of a theory. If we allow for more than one member of each $I_j$ to be considered in the empirical claim we arrive at the following.

<u>D6</u> If $T = \langle M_p, M, M_{pp}, I, (I_j)_{j\in J}, (\mathcal{U}_j)_{j\in J}\rangle$ is a theory in the sense of D4) then

a) the <u>empirical claim of</u> T <u>with precision</u> $(U_j)_{j\in J}$ ($U_j \in \mathcal{U}_j$ for all $j \in J$) is that

$$\exists \mathfrak{X}(\mathfrak{X}\in \prod_{j\in J} Pot(I_j) \wedge \bar{\rho}(\mathfrak{X}) \subset r(M) \wedge \forall j\in J(\mathfrak{X}(j)\neq\emptyset \wedge (\mathfrak{X}(j)\times\mathfrak{X}(j))\subset U_j))$$

b) the <u>vague empirical claim</u> of T is that for each $j \in J$ there is exactly one $U_j \in \mathcal{U}_j$ such that

$$\exists \mathfrak{X}(\mathfrak{X}\in \prod_{j\in J} Pot(I_j) \wedge \bar{\rho}(\mathfrak{X}) \subset r(M) \wedge \forall j\in J(\mathfrak{X}(j)\neq\emptyset \wedge (\mathfrak{X}(j)\times\mathfrak{X}(j))\subset U_j))$$

According to D6-a) for every concrete system $j \in J$ there are some members of $I_j$ (at least one, since $\mathfrak{X}(j) \neq \emptyset$) which can be extended to models <u>and</u> which are similar to each other with degree $U_j$. D6-b) makes sense only in view of our use of <u>weak</u> uniform spaces. For with full uniform spaces $\mathcal{U}_j$ a claim of this form would always be true. Existential quantification over the $U_j$ becomes non-trivial only if the $\mathcal{U}_j$ are obtained by 'cutting off' full uniformities on $I_j \times I_j$ at some place 'in the direction of dissimilarity'. We then can imagine one or several 'greatest' degrees of similarity in $\mathcal{U}_j$ which already are 'fine' enough to express real, non-trivial similarities.

To sum up this section we can say that the kind of vagueness of empirical claims discussed here is due to a statistical character of the world. Every measurement reveals a slightly different object which is measured. So it is misleading to say one has measured one object and obtained different results because of 'disturbances' or what ever. It is more adequate to say that by repeated (eventually sharp) measurement one has acquired information about a multiplicity of slightly different objects. This kind of vagueness has to be considered as a component of empirical theories because it affects the empirical claim in a straightforward way. It corresponds to Ludwig's 'unscharfen Abbildungsprinzipien' and to Moulines' uniformities on the non-theoretical level. My proposal to take the content of 'vagueness a multiplicity' seriously is to consider a new kind of empirical claim which is just an extended Ramsey-sentence.

## III. VAGUENESS AND IDEALIZATION

A second kind of vagueness comes up by the observation that our theories and theoretical pictures are 'idealizations', whereas the world is not idealized, i.e. 'real'. In an empirical claim these two things have to be put together. But they are of different nature and so how can we manage to bring them together? Since both parts do not precisely fit with each other, some room for vagueness seems necessary.

In Sneed's theory concept this problem does not exist at first glance. For intended applications, which represent 'the world' in the empirical claim already have an 'ideal' mathematical structure, namely that of partial potential models. In case of CPM, for instance, partial potential models are entities $\langle P,T,s\rangle$ which contain a continuous position function s. So the entities to be combined in the empirical claim, namely I and M, are of the same idealized nature. But this answer clearly is not satisfactory. The problem is still present in the connection between I and the world.

The contrast between idealized pictures and real world is clearly present in the relation between intended applications and concrete physical systems. The kind of vagueness we are thinking of is therefore implicit in the problem of how intended applications are determined (by means of measurement). For theories whose partial potential models contain quantitative components this amounts to the question: Can the set of possible outcomes of our measuring instruments include the set of all real members? Clearly the answer is: No. The number of distinguishable outcomes is usually rather small. This certainly does not put us in a position to read off the function values of a continous quantity for all its arguments from the real system. The question then is how to bridge the gap between our crude 'observational data' read off from measuring instruments and the idealized continuous quantities occurring in the theory.

Before continuing here it seems helpful to see how this question is related to the kind of vagueness discussed in Sec. II. There we have presupposed that 'by means of measurement' we somehow arrive at sharp values. This assumption seems to be false in the light of the present considerations. But this does not detract from the results of Sec. II, for what has been said there does not depend on whether our theories are idealized. The basic feature of 'vagueness as multiplicity' is present also in cases of non-ideal theoretical pictures. So it seems that 'vagueness as multiplicity' is independent from 'vagueness induced by idealization' (if something like that exists). Whether the opposite also is true depends on what we can make out of the latter kind of vagueness. In any case the discussion of this section does not affect the results of Sec. II. Even if we find it difficult to understand how we can arrive at 'sharp' results from 'coarse' data, this is no reason to attack 'vagueness as multiplicity'. For as long as we have our idealized theories, and no better ones, we shall want to work with these and somehow to bridge the gap, even if we do not quite understand how. And as soon as we have such a bridge 'vagueness as multiplicity' becomes relevant.

To come back to the question whether we need some formal representation of the vagueness arising because of the gap between real world and idealized theories, let us consider the proposals to bridge this gap. A first proposal which treats intended applications in the form of atomic obvervational propositions is to substitute propositions of the form '$f(x) = \alpha$' by '$\alpha - \varepsilon < f(x) < \alpha + \varepsilon$' where f is a physical quantity having real numbers as values. This account seems to be very natural and in fact has been proposed by different authors, e.g. Ludwig [7] and Wojcicki [16]. It is natural because it reflects physical practice: whenever in reading off some value from an instrument we stop at a certain decimal we have done something like introducing such an interval in which the 'real' value has to lie.

Even if this proposal is satisfactory to some extent as a description of practice it does not yield a better, in the sense of 'more precise' unterstanding of how we bridge the gap between reality and theory. On the one hand observational results of the form '$\alpha - \varepsilon < f(x) < \alpha + \varepsilon$' still are 'sharp' in the sense that they contain special real numbers, namely $\alpha - \varepsilon$ and $\alpha + \varepsilon$. These numbers are not given by experience, they rather are chosen according to some pragmatic standard we usually are not able to make explicit. But why not use similar standards to fix a unique value $\alpha$ thus obtaining propositions of the original form

'$f(x) = \alpha$'? The only difference between these two versions is that the 'interval'-version contains an explicit hint that the observational proposition contains some conventional component, namely in the choice of $\varepsilon$; while the original version tends to make us forget this fact. From a formal point of view I see no reason to say that the 'interval'-version is 'nearer to reality' than its alternative.

On the other hand the question arises how to formulate an empirical claim with intended applications using such 'interval-statements'. It has been proposed to consider a physical theory as 'useful' if the observational sentences added to the axioms of the theory do not lead to contradictions [7]. In terms of set theoretic structures this amounts to a rather weak Ramsey-claim, namely that it is possible to 'fill in', first, the gaps between the finitely many observational propositions in order to obtain continuous functions, and, second, to add theoretical components in order to obtain full models. If we regard intended applications as full partial structures then an analogue to such a claim could be formulated as follows. We first smear the set of originally given sharp intended applications by introducing some uniformity at the non-theoretical level. But of course we cannot choose I as the basic set of the uniform space for we want to have included systems similar to, but different from intended applications. Therefore we have to choose a set of structures containing I as the basic set of the uniform space. A natural candidate for this is the class of all partial potential models, $M_{pp}$. By introducing a uniformity on $M_{pp}$ we can formulate a claim of the following form. For each intended applicaiton $x \in I$ there exists a partial potential model x' in some neighbourhood U of x such that x' can be extended by theoretical components to a model. More formally:

$$\exists X(\forall x \in I \exists ! x' \in X(\langle x,x'\rangle \in U) \wedge X \subset r(M))$$

where U is a fixed degree of similarity of the uniformity imposed on $M_{pp}$. This claim can be reformulated as follows. We denote by $U_x$ the neighbourhood of x induced by U, i.e. $U_x = \{x'/\langle x,x'\rangle \in U\}$. Then the above formula is equivalent to

$$\exists X(X \in \prod_{x \in I} U_x \wedge \rho(X) \subset r(M))$$

There is an analogy between such a claim and propositions of the form '$\alpha - \varepsilon < f(x) < \alpha + \varepsilon$': I corresponds to the sharp $\alpha$ and U to $\varepsilon$.

But if we compare this approach with that of D5) we see that there is not much difference, at least formally. In fact, it just amounts to the possibility we already mentioned in the discussion of D5). The only difference is that in D5) the basic set of the uniform space is given by I where I is different from $M_{pp}$, while here we have no clear idea how to choose this basic set (hence the idea to take all of $M_{pp}$). But the disadvantage of this technical formulation has already been pointed out: it suggests that there is something like sharp, real intended applications. We see no argument for such a thing to exist and therefore no argument to prefer such a formulation to the more neutral one of D5).

So the proposal to go over to 'smeared' intended applications in order to bridge the gap between reality and theory, as far as it formally enters into the theory concept by imposing a uniformity on the intended applications, adds nothing to D5) and thus is redundant. The mechanism of choosing a special $\varepsilon$ or a special U of some uniformity at present is not theoretically understood and our understanding is not much improved by pointing out that we choose some such $\varepsilon$ or U. I do not feel convinced that this formal treatment gives insight into the mechanism.

A second proposal is due to Suppes [15]. But his work contains more of a program than a clear picture of how things are. Although I think that a theoretical understanding in fact can be achieved in the direction Suppes indicates, I have nothing to add at the moment.

We will have to live at the moment with the fact that we do not really know how 'observations' can be brought together with theories. Of course it is possible to develop theoretical pictures of such a mechanism but, I think it would be more efficient to study in detail some concrete examples first. Let me note that here we have another example of the 'openess' of Sneed's theory concept. In order to understand or to develop a theoretical picture of how 'data' come into contact with theories, we do not need to change the theory concept at all. For such a theoretical picture can be developed by means of putting together theories of the form of D2) (or D4)) in special ways. A special relation between theories (in the sense of D2)) which seems to be relevant here is 'theoretization' as introduced in [1] and [3]. But perhaps we need further relations not yet investigated.

Let me close with a more philosophical aspect of this problem. One clear feature of idealization is that some idealized theories contain an 'induction along space' in both directions: towards infinitly big and small areas. This is true for all theories containing Euclidean geometry or the real numbers ('containing' here in the logical sense). The axioms of such theories have been confirmed with objects of a certain size. But the way these theories are axiomatized induces their models to be infinite and to be infinitely fine (continuous). The use of such models implicitly assumes that things in certain respects behave independently of their size. And to a certain extent this assumption is justified by a form of induction. We assume that in realms in which we do not yet have experience, the world is like in that area we know. But of course such an assumption can be, and in fact has been, questioned. Today we do not believe in that induction. So one way to bridge the gap between data and theories, at least with respect to this special type of idealization, would be to adjust our theories. One could try to eliminate such idealizing assumptions as far as possible and thus obtain theories whose basic terms can already be interpreted as 'real' objects in some specified domain. H.-J. Schmidt's contribution to this volume is one step in that direction. But such modification of physical theories is physics and does not belong to philosophy of science.

BIBLIOGRAPHY

[1] Balzer, W., Empirische Geometrie und Raum-Zeit-Theorie in mengentheoretischer Darstellung, Kronberg i.Ts. 1978
[2] Balzer, W. and Moulines, C.U., On Theoreticity, Synthese 44 (1980)
[3] Balzer, W. and Sneed, J.D., Generalized Net Structures of Empirical Theories, Studia Logica 36 (1977) and 37 (1978)
[4] Bourbaki, N., Theory of Sets, Paris 1968
[5] Feferman, S., Two Notes on Abstract Model Theory, 1, Fund. Math., 82, 1974
[6] Ludwig, G., Deutung des Begriffs "physikalische Theorie" und axiomatische Grundlegung der Hilbertraumstruktur der Quantenmechanik durch Hauptsätze des Messens, Lecture Notes in Physics, 4, 1970
[7] Ludwig, G., Die Grundstrukturen einer physikalischen Theorie, Berlin-Heidelberg-New York 1978
[8] Moulines, C.U., Approximate Application of Empirical Theories: A general Explication, Erkenntnis 10 (1976)
[9] Przelecki, M., The Logic of Empirical Theories, London 1969
[10] Przelecki, M., Fuzziness as Multiplicity, Erkenntnis 10 (1976)
[11] Przelecki, M., Some Approach to Inexact Measurement, Poznan Studies in the Philosophy of Science and the Humanities 4, (1978)
[12] Sneed, J.D., The Logical Structure of Mathematical Physics, Dordrecht 1971
[13] Stegmüller, W., The Structure and Dynamics of Theories New York-Heidelberg-Berlin 1976
[14] Stegmüller, W., The Structuralist View of Theories, Berlin-Heidelberg-New York 1979
[15] Suppes, P., Models of Data, in: Logic, Methodology and the Philosophy of Science (E. Nagel et al., eds.), Stanford 1962
[16] Wojcicki, R., Basic Concepts of Formal Methodology of Empirical Sciences, in: Twenty-five Years of Logical Methodology in Poland (M. Przelecki and R. Wojcicki, eds.), Dordrecht 1977

MYSTICAL REALISM

A de-Sneedified Program of Formalization
and an Exercise in Einsteinian Methodology

R.M. Cooke

Technische Hogeschool Delft
Kanaalweg 2b
2600 GB Delft
Netherlands

## INTRODUCTION

## A DE-SNEEDIFIED PROGRAM OF FORMALIZATION

Two things strike me in contemporary discussions about theoreticity. First, in those studies of individual theories known to me, the number of theoretical terms is the same, namely two [14, 16]. Second, in all the abstract treatments of theoreticity, the theoretical terms are indexed from r to n ($r < n$). Are there always only two theoretical terms? If not, what happens when the theoretical vocabulary gets largs, is it simply a matter of letting an index run up to n? I think the answer to both questions is "no".

When Riemann read his paper "On the Hypotheses which Lie at the Foundations of Geometry" in 1854, it became received opinion that "Either ... the actual things forming the groundwork of a space must constitute a discrete manifold, or else the basis of metric relations must be sought for outside that actuality, in colligating forces that operate upon it" [20 p. 425]. In the classic foundational studies of Mach [11] and Hertz [8], we find no explicit attempt to "derive geometry from mechanics" but we do find a recognition that geometry and chronometry are not simply empirical sciences.

Sneed [16] establishes without too much difficulty the claim that there is no theory-free method of measuring force. When he makes a similar claim for mass, he must consider the follwing possibility: "It is often suggested, for example, that 'mass' might simply be "defined" as that quantity determined by using an analytical balance in the customary way. This way of determining mass values would not then presuppose the truth of <u>any</u> physical laws". He blocks this possibility by remarking: "When we claim that an analytical balance determines mass-ratios accurately, we are claiming, at least, that there is a way of regarding this physical system as a model for CPM" [16 p. 117]. However, similar statements certainly hold for measurements of time intervals by means of a standard clock, and for measurements of space intervals by means of a rigid body. His argument for the theoreticity of mass would also estabilsh the theoreticity of the metrical concepts for space and time. (Sneed himself conceeded this in a discussion following a lecture at the T.H. Delft, December 1975.)

The Sneed formalism is well suited to capture what physicists 'have in mind' when they do classical particle mechanics (CPM). However, I believe that it is not well suited for recognizing the full scope of theory-dependence, or for analysing its methodological consequences.

The methodological consequences of theory-dependence can best be appreciated on the basis of an analysis of content. Content is a semantical notion. Since the standard Tarski semantics of the first order predicate calculus is better understood than the semantics of the set-theory formalism, I have chosen to formalize CPM in a many-sorted first order language (Appendix I). The paper itself discusses the notion of content in relation to methodology.

AN EXERCISE IN EINSTEINIAN METHODOLOGY

Einstein's methodological position in, for example, his Herbert Spencer Lecture of 1933, may be described as *mystical realism:*

> If, then, it is true that this axiomatic basis of theoretical physics cannot be extracted from experience but must be freely invented, can we ever hope to find the right way? Nay more, has this right way any existence outside our illusions? ...I answer without hesitation that there is, in my opinion, a right way, and that we are capable of finding it. Our experience hitherto justifies us in believing that nature is the realisation of the simplest conceivable mathematical ideas. I *am convinced that we can discover by means of purely mathematical constructions the concepts and the laws connecting them with each other, which furnish the key to the understanding of natural phenomena.* ([5], p. 136)(emphasis added).

This position is articulated (and not, as some suggest, retracted) in his *Autobiographical Notes*. Here Einstein distinguishes two points of view for judging physical theories, namely, external confirmation and inner perfection. Before embarking upon an 'interior critique' of classical mechanics (pointing up its inner imperfections) he remarks "...such a critique is well suited to show the type of argument which, in the choice of theories in the future will have to play an all the greater role the more the basic concepts and axioms distance themselves from what is directly observable, so that the confrontation of the implications of theory by the facts becomes constantly more difficult and more drawn out" ([6], p. 27).

Einstein suggests that methodology should possess an historical aspect. When a science is in its infancy, the criterion of external confirmation swamps the criterion of inner perfection in determining rational theory choice. As a science becomes more mature, the criterion of inner perfection becomes more important and eventually dominates. We do not yet know how this process ends. This is a question for the history of science, not for the methodology of science; and the history of science is not yet over.

Einstein's methodology seems to contain a paradox: On the one hand he claims that the process of maturation for a science is a process of approaching, and eventually grasping the real. On the other hand, the method according to which the more mature theories are selected appeals <u>less</u> to external confirmation and <u>more</u> to the internal properties of the theories themselves.

I speculate that it was Einstein's "philosophical intuition" which led him to entertain both of these beliefs, yet he was never certain whether this combination of beliefs could be shown consistent. Perhaps this is the reason he describes himself as an opportunist in matters philosophical ([6], p. 684).

I want to suggest that these beliefs, <u>can</u> be consistently combined. Moreover, a methodology which consistently combines them is, in a certain sense, the historical result of the methodology appropriate for late-classical mechanics. By late-classical mechanics I mean the theory whose foundations were investigated by Mach [11] and Hertz [8]. In retrospect, the foundational problem with which these men grappled arose from the recognition of one or two theoretical terms. The methodology for a science which sees itself in this situation is exactly the methodology which Sneed so elegantly formalized. Based on his experience with general relativity, Einstein saw that theoreticity must become a more pervasive phenomenon in theories, and must come to play a role essentially different from the role it plays in the methodology of late classical mechanics (i.e. Mach's positivism).

This expanded role of theoreticity is already implicitly present in late classical mechanics. Hopefully, the formalization presented here (Appendix I) provides a framework for recognizing this role and analysing its methodological consequence. While I cannot claim that Einstein's methodology becomes less mystical on this analysis, I do think it becomes less opportunistic.

## I

A formalization of an informal theory $T^o$ is a map from $T^o$ into a theory T in some formal language. The mapping $T^o \to T$ will preserve some features of $T^o$, and we choose the mapping in such a way that it preserves those features which we wish to study. In this section we define a set of features which we wish to preserve and study in a formalized theory. By a semantic interpretation of a term in an informal theory we shall understand simply a rule assigning an extension to the term. The only restriction which we place on this rule is that it be invariant with respect to the theory in which the term in considered. The same rule must apply regardless which theory we choose to consider; indeed, if this were not so we should hardly speak of an interpretation for a term. Of course, a term can get reinterpreted according to the interests of some theory. In this case we speak of a theory-dependent choice of semantic interpretation, and we say that the term is normed to the theory.

D0. A semantic interpretation for a term in an informal theory $T^o$ is a rule assigning an extension to the term.

D1. A semantic interpretation of a term is complete if its rule is decidable, that is, if given any argument there is an effective procedure to decide whether the argument belongs to the term's extension.

D2. A term's semantic interpretation is rigidly attached with respect to $T^o$ if the non-validity of $T^o$ in any model in which the term has the specified interpretation is not a sufficient motive for altering the term's interpretation. In this case we shall say that the term has a rigid semantic interpretation with respect to $T^o$, or simply that it is rigid with respect to $T^o$. A term is rigid if it is rigid with respect to every theory in which it occurs.

D3. A term in $T^o$ is objective (with respect to $T^o$) if its semantic interpretation is rigid (with respect to $T^o$) and complete. An objective term is called an O-term.

D4. A term in $T^o$ is theoretical-interpreted with respect to $T^o$ if its semantic interpretation is complete, but chosen in a theory-dependent way; that is, if a change in the semantic interpretation of this term is adequately motivated whenever such a change removes a discrepancy between the theory and the interpretation of its terms.

D5. A term in $T^o$ is theoretical-non-interpreted with respect to $T^o$ (TN) if it is without a semantic interpretation.

D6. An informal theory is regular it its non-logical vocabulary constists of O, TI and TN terms only. (The vocabulary of set theory is considered part of the logical vocabulary.)

Absolute rigidity and completeness of semantic interpretation are of course idealizations. In concentrating on regular theories we adopt these idealizations, and it is therefore appropriate to justify them. In doing foundations we are naturally interested in the most fundamental theories available. A fundamental theory may be defined as one whose non-rigid vocabulary is normed to the theory itself. Hence the requirement that the non-logical vocabulary consist only of O, TI and TN terms entails that the theory in question is fundamental. A fundamental theory can always be gotten by taking the conjunction of non-fundamental ones (provided these are consitent).

I oppose the use of any "operational hierarchies" [9] in order to solve methodological problems, especially problems involving theory-dependence. For one thing, such a fragmentation of physics is contrary to the drive for unified theories. For another thing, such hierarchies seems to create knowledge (e.g. about univocal values for theory-dependent functions) which doesn't really exist.

Complete semantic interpretations are also idealizations. Most terms in a physical theory are not provided with a complete semantic interpretation, as Carnap showed in his classic [1]. But a partially interpreted term cannot be handled in a formal language with a nice semantics. Sharp semantic concepts become applicable if the terms in a physical theory are classified according to the limiting distinction: complete semantic interpretation / no semantic interpretation.

This distinction, however, does not coincide with the distinction: rigid / non-rigid, and multiplying these two distinctions, a two-by-two matrix is generated in which the entry rigid-and-non-interpreted is empty by definition. For a regular theory, then we may fill in the matrix as shown below:

| | complete semantic interpretation | no semantic interpretation |
|---|---|---|
| rigid | O | - |
| non-rigid | TI | TN |

Significant here is that the theoretical vocabulary splits, and this derives from the fact that the two defining properties for O-terms, rigidity and completeness, can fail independently.

The distinction: theoretical-interpreted / theoretical-non-interpreted can be filled in as follows. Some terms are completely operationalized, but in a way which strongly depends upon some background theory. In the social sciences this phenomenon is perhaps better knows, although the background theory is generally less clear than in the natural sciences. Terms like "intelligence" and "efficiency" are operationalized by means of tests, which themselves are continuously being tested against background theoretical assumptions.

A theory with TI terms can conflict with its interpretation, that is, with the interpretation of its terms. However, whenever possible, the theory would exercise its normative mandate and prescribe a re-interpretation of TI terms in order to produce an interpretation with which it agreed.

The TN terms, on the other hand are not operationalized at all. There is not theory-external means for assigning them an extension. If the remaining terms of the theory are interpreted, then it may be possible to assign an extension to the TN terms in such a way that a model for the theory is produced, but this possibility need not exist. We cannot speak here of a semantic interpretation in the sense introduced at the begining of this section. Of course, in a formalized theory we can always assign an arbitrary interpretation to a TN term, but this does not blur the distinction with respect to the informal theory.

It must be emphasized that these distinctions contain a strong pragmatic component; they distinguish different types of behavior which scientists exhibit with respect to various terms. This is not the place to discuss whether they contain other components as well.[1)]

## II

We switch now to formalized theories, that is, theories in some formal language. A formalization of CPM in the first order predicate calculus, preserving the distinctions of the previous section, is given in Appendix I, and is summarized briefly in section III. In this section, let T be a formal representation in a first order language of a regular informal theory $T^o$. We may decompose the vocabulary $V_T$ of T as follows:

$$V_T = V_L \cup V_O \cup V_{TI} \cup V_{TN}.$$

Let $L_L$ be the language generated by the logical vocabulary $V_L$ (including the vocabulary of set theory): let $L_O$ be the language generated by $V_L \cup V_O$; $L_{TI}$, by $V_L \cup V_O \cup V_{TI}$ and let $L_T = L_{TN}$ be generated by $V_T$.

A structure for $L_T$ will be denoted:

$$M = \langle |M|, L^M : O^M : TI^M : TN^M \rangle$$

where $|M|$ is a domain of elements, and $L^M$, $O^M$, $TI^M$, $TN^M$, are interpretations of the L, O, TI and TN vocabularies in the domain $|M|$.

Before defining content we must first define intended interpretations for $L_T$.

D7. An intended interpretation of $L_O$ is a structure for $L_O$ in which the $V_L$ terms receive a standard mathematical interpretation, and in which the $V_O$ terms are interpreted according to the rigid rules by virtue of which they belong to $L_O$. An intended interpretation of $L_T$ is an intended interpretation of $L_O$ in which the theoretical vocabulary has been interpreted.

This definition entails that the terms in $V_O$ must receive a physical interpretation, they cannot be interpreted in the mathematical universe. We shall write:

$$T = T_N \supset T_{TI} \supset T_O \supset T_L;$$

where $T_{TI}$ are the axioms of T which contain no TN terms, etc.

D8. An intended model of T is an intended model of $L_T$ in which T holds.

D9. If $M = \langle |M| L^M : O^M \rangle$ is a model of $L_O$, then $M' = \langle |M| L^M : O^M : TI^M : TN^M \rangle$ is an expansion of M to a structure for the language $L_T$.

The content of a physical theory is intended as a measure for its informativeness with respect to the intended interpretations of the language in which the theory is expressed. If a theory holds in all intended interpretations of the formal language, then it is quite uninteresting and deserves a low grade of content. Conversely, if it holds in only a small class of all possible intended interpretations of its language, then it tells us quite a lot about the interpretations in which it holds. Content thus corresponds to excluded possibilities. (In this sense a contradiction has maximal content; an alarming consequence at first sight but nevertheless correct. If one knows that a contradiction is the case, then one knows all there is to know.)

D10. The objective content of T (Obct(T)) is the class of conceivable intended interpretations of $L_O$ which cannot be expanded to models of T.

D11. The theoretical content of T (Thct(T)) is the class of conceivable intended interpretations of $L_{TI}$ which cannot be expanded to models of T.

We must direct our attention to all conceivable intended interpretations because we are working with physical interpretations, and if the theory holds generally in our world, then all applications of the rigid semantical rules will yield (intended) models which can be expanded to models of the theory.

The formulation of definitions D10 and D11 is a bit unsatisfying, for the notion of a conceivable intended interpretation is rather vague. We may think of D10 and D11 as defining absolute notions of objective and theoretical content, for these are in fact limit concepts which we must approach through other concepts. The problem is to restrict the class of e.g. conceivable results of applying the semantic rules for $V_O$ to a class with manageable properties. The way to achieve this is to define suitable concepts of relative objective and theoretical content.

The proper treatment of relative content requires some caution. Hempel has made a proposal which we might consider in the present context:

> Now, the common content of two statements is expressed by their disjunction, which is the strongest statement logically implied by each of them. Hence, the common content of h and k is given by $h \vee k$. But h is logically equivalent to $(h \vee k) \wedge (h \vee -k)$, where the two component sentences in parentheses have no common content: their disjunction is a logical truth. Hence that part of the content of h which goes beyond the information contained in k is expressed by $(h \vee -k)$. ([7], p. 154) notation changed slightly).

In order to evaluate this proposal we need the following theorem. If a is a sentence in $L_T$, we put: obct(a) := $(M \in MOD(L_O) \mid$ no expansion $M' \in MOD(L_T)$ of M satisfies a). "-obct(a)" denotes the subclass of models $M \in MOD(L_O)$ which do not belong to obct(a). The dependence of obct(a) on T is ignored in this notation.

THEOREM 1

If a, b, sentences in $L_T$, then

(i) $\text{obct}(-a) \subset -\text{obct}(a)$

(ii) $\text{obct}(a) \cup \text{obct}(b) \subset \text{obct}(a \wedge b) \subset -\text{obct}(-a) \cup -\text{obct}(-b)$

(iii) $\text{obct}(a \vee b) = \text{obct}(a) \cap \text{obct}(b)$

*Proof*. Straightforward substitution from D10.

Similar relations hold for theoretical content.

D12. The <u>objective content of a relative to be in the sense of Hempel</u> is $\text{obct}(a \vee -b)$.

This definition has at least one obvious shortcoming, if $\text{obct}(-b) = \emptyset$, or if $\text{obct}(a) = \emptyset$, then by theorem 1 (iii) $\text{obct}(a \vee -b) = \emptyset$. However, it is quite possible in this situation that $\text{obct}(a \wedge b) \neq \emptyset$.

D13. The <u>objective (theoretical) content of a relative to b</u> is obct(a∧b) - obct(b) (thct(a∧b) - thct(b)).

THEOREM 2

The objective content of a relative to b contains the objective content of a relative to b in the sense of Hempel.

*Proof*. Using Theorem 1:

$$\begin{aligned} \text{obct}(a \wedge b) - \text{obct}(b) &\supset \text{obct}(a) \cup \text{obct}(b) - \text{obct}(b) \\ &\supset (\text{obct}(a) \cup \text{obct}(b)) \cap \text{obct}(-b) \\ &= \text{obct}(a) \cap \text{obct}(-b) = \text{obct}(a \vee -b). \end{aligned}$$

We may now reason as follows. Normally we will be able to formulate axioms in $L_O$ which describe the interesting applications of the semantic rules for the objective vocabulary. If for example $L_O$ contains the term "spatial point", then we might have an axiom saying that the set of spatial points can be mapped bijectively into $R^3$. If $T_O$ contains only axioms of this sort, then we may consider the objective content of T relative to $T_O$ as a reasonable approximation of the absolute objective content. Similar remarks apply for the theoretical content. Now we can study and resolve a relative content question, and, if desired, discuss the question whether we have an adequate approximation of the absolute content before us.

## III

To illustrate these notions we summarize the treatment of classical particle mechanics given in Appendix I. The intended interpretations of the language of CPM are finite sets of mass points moving moving in space and time. The objective vocabulary is taken to consist of space, time and mass points. That is, we assume that we possess effective procedures for deciding whether something is a space - a time - or a mass point. With regard to space and time this is given by a coordination procedure assigning numbers or number triples to points of time and space respectively. [N.B. by separating the domains of space and time points in this way we admit space and time points as absolute. Physically this is a matter of some delicacy of course, but it is logically more perspicuous to proceed in this way in an initial treatment.] Furthermore, these effective procedures are rigid with respect to CPM. Although we may adopt a new coordination system, the set of points which gets coordinated remains the same; we are not prepared to revise our thinking on this score in order to produce a better fit between CPM and its interpretation.

Regarding mass, it is fixed beforehand what sorts of things receive mass values; that is we possess an effective procedure to determine whether an object belongs to the class of things describable by mechanics. For example, rocks belong to this class, but shadows and temperature zones do not, even though these latter can also undergo displacements in space and time. We do not claim to know before hand which mass values the members of this class receive, only that they receive some value or other (greater than zero).

It is a different matter with the metrics which we place on the sets of mass, space and time points. In operationalizing these metrical concepts we frequently appeal directly to the theory for guidance. Metrical concepts are also completely interpreted, at least in principle. Once the appropriate measurement procedures are chosen, we can in principle measure every space and time interval, and measure the mass of every body. Of course this is not true in quantum mechanics, but on the horizon of classical mechanics it is held to be unproblematical. These metrical concepts belong therefore to the TI vocabulary.

There remains the notorious term "force". Force is the theoretical term par excellence; in the International Bureau of Weights and Measures in Sevres, France, there is no box in which the standard force is kept, for force is not interpreted by direct operationalization. The classic foundation studies of Mach [11] and Hertz [8] were directed to the problem of formulating mechanics in such a way that force does not appear as a primitive term. It was to be admitted only as a term defined by the relation $F = ma$.

## IV

I claim that the above classification is more or less implicit in the foundational studies of late classical mechanics. A few remarks are in order.

In the first place, the theoretical status of the metrical concepts for space and time is not explicitly acknowledged. Hertz, for example, introduces Euclidean geometry as a priori. However, he is visibly embarrassed with this move and indicates that he is also prepared to regard it as a convention.

The case of Mach is also interesting. On the basis of his critique of absolute space and time, he drew the conclusion that the metrics for space and time were also not absolute. He considered them arbitrary conventions which we choose according to our intuitive feeling for spatial and temporal equality. This is rather bizarre, for his definition of mass can lead to the values zero and infinity if these conventions are chose in the wrong way.[2] In this case the mass ratio will not be defined between all pairs of bodies. Mach cannot leave everything to convention. On the other hand, Mach could not derive metrics for space and time from the laws of motion (as he did for mass-ratios) because the laws of motion would not lead to a unique choice. Classical mechanics needs external operalization for at least some of its metrical concepts.

The manner in which the metrical concepts for space and time hover between the theoretical and non-theoretical vocabularies in Mach's mechanics shows the need for a more nuanced treatment of theoreticity. With the distinction between interpreted and non-interpreted theoretical terms, we can regard the choice of a metric as fully theory depended, without requiring that a unique metric be derivable from the theory. In this case the theory has veto powers, but no power of appointment. Only in this way can we do justice to the insights of Gauss and Riemann into the relation between geometry and physics (recall, these insights were made with respect to classical mechanics).

Mach proposed to derive mass values from the laws of motion in combination with space and time measurements, a procedure appropriate for TN terms. Hertz, on the other hand, placed the measurement of mass on the same level as the measurement of space and time. However, he admitted that mass points could be introduced by hypothesis and their values calculated via the laws of motion. There is no consensus regarding the status of mass in late classical mechanics. I have chosen to classify "mass point" as an O term because (a) even Hertz would acknowledge that part of this term's extension is rigidly attached, and (b) Hertz' hypothetical masses never gained wide acceptance. I have chosen to classify "mass value" as a TI term because (a) it admits operationalization in terms of a balance, and (b) Mach's method of deriving mass values works only for small systems (cf. [16]).

## V

Regarding the objective and theoretical content of CPM, some results are developed in Appendix II. For the present we can be brief. A conceivable system composed of a finite set of mass points moving around in space and time belongs to obct(CPM) just in case there is no possible choice of a metric for space, time and mass, and no force function under which this system conforms to the laws of motion. Obct(CPM) is not empty, but it is not very large, and membership in this class is not very interesting.

A conceivable intended interpretation of $L_{TI}$ is a finite set of mass points moving in space and time, where space and time are provided with metrics, and mass values have been assigned. These metrics need not be the familiar ones. Such a system belongs to thct(CPM) just in case there is no force function which would bring the system into agreement with the laws of motion. In the special case where the familiar metrics for space and time are assigned, the question whether a given system belongs to the theoretical content class of CPM would count as a meaningful and interesting question. If a late-classical physicist was asked, what does CPM say about our world W, he might well answer (in effect): "It says $W \notin$ thct(CPM)". Since thct(CPM) is rather large, this would indeed be significant and informative. It would not be particularly significant to learn that $W \notin$ obct(CPM).

## VI

We come now to an interesting philosophical juncture. The people who did foundational research in late classical mechanics did not believe in the existence of a "true" clock or a "true" meterstick. For them such expressions were meaningless ... their positivist theory of meaning told them that meaning was method of verification, and there was no way of verifying propositions like "X is the true meterstick". The choice of a clock or a meterstick was purely a matter of convention; ontologically speaking, any choice would be as good as another. Someone who believes this must admit that the question, 'does our world W belong to thct(CPM)?' is not convention-free. Under some measuring conventions it may belong (i.e. there is no force function which makes W a model for CPM under these conventions) and under other conventions it may not belong (i.e. there is such a force function). Hence, if such a person really wants to know what CPM says about the world, then 'is $W \in$ thct(CPM)?' is simply the wrong question.

Why were late classical physicists asking the wrong question? Why did the workers in foundations mention (with embarrassment) the theory-dependent character of the metrics of time and space, only to forget about it as soon as possible? In light of what was said about obct(CPM) in the previous section, one answer may be this: the question 'Is $W \in$ obct(CPM)?' is *also* the wrong question, although it is wrong for a different reason. If CPM's convention-free information about the world W is captured in the sentence $W \notin$ obct(CPM) (i.e. our world obeys the laws of mechanics, under some appropriate measuring conventions, and under some appropriate force function), then CPM has precious little to say.

If our methodology is leading us to ask wrong questions, it may be time to consider changing methodologies.

## VII

I mentioned in the Introduction that Einstein's methodology presents us with a paradoxical aspect. It combines a realist posture with respect to the theoretical vocabulary and an idealist posture with respect to the criterion of rational theory choice, for mature theories. By way of conclusion I would like to give a sort of "methodological derivation" of Einstein's position. I certainly do not claim that this derivation was actually used by Einstein. In physics it is often useful to be able to derive a law in several different ways; perhaps the same is true in methodology.

The methodological malaise of late classical mechanics may be put succinctly as follows. Positivism teaches that only the $L_O$ language is capable of describing the real physical world. The rest is conventions and mathematical formalism. However, strict application of the positivist criterion of meaning tends to make $L_O$ smaller, and this decreases our theories' expressive power with respect to the real world (assuming that expressive power is understood in terms of objective content, as indicated in section II). In late classical mechanics these opposing demands begin to make themselves felt.

For a positivist, the theory-dependent character of theoretical terms derives from the positivist criterion of meaning. There _is_ no such thing as force, there _is_ no such thing as a true meterstick, etc. Theories have the mandate to prescribe interpretations for these terms because they lack independent (positivist) meaning. If it's all a matter of convenience, we may as well let our theory decide what is most convenient. I call this the _conventionalist reading_ of the problem of theoretical terms.

A reform in methodology will require a new explanation of theory-dependence. Instead of explaining theory-dependence in the positivist fashion, we could say the following: for a given interpretation of the O vocabulary, there _is_ a correct interpretation for the theoretical vocabulary. We would say, there _is_ such a thing as force, there _is_ such a thing as mass (in the sense of quantity of matter), and there _is_ such a thing as spatial distance, etc. The normative mandate which a theory enjoys in prescribing interpretions for these terms comes from the fact that the theory is the best, or perhaps the only, instrument for discovering what these true interpretations are. I call this the _ontological reading_ of the problem of theoretical terms.

Someone wishing to escape the methodological malaise of late classical mechanics will be motivated to abandon the conventional reading and adopt the ontological reading of the problem of theoretical terms. This constitutes a "derivation" of the realist posture.[4]

Now suppose that this person recognized that the normative mandate of a theory is not confined to the TN terms but extends to other terms which are operationalized in a theory-dependent manner (i.e. TI terms). As the case of CPM suggests, it is reasonable to suppose that the objective content of the theory becomes "small" when the theoretical vocabulary is enlarged in this way. This means that "almost any" model of the non-theoretical vocabulary will be expandable to a model for the theory. Let the theory in question be T, and let T' be a competing theory. In order for T and T' to compete, they must be about the same things, that is, they must have the same O vocabularies and the same rigid interpretations for the O terms. Without losing generality we may suppose that their theoretical vocabularies also coincide (this can be arranged by adding superfluous terms to the languages of the two theories). According to the above supposition, both obct(T) and obct(T') will be "small".

Now what would it mean to decide between T and T' on the basis of external confirmation? It would mean that our world was found to belong to obct(T) but not to obct(T'), or vice versa. However, the case that $W \in$ obct(T) or $W \in$ obct(T') will "almost never" arise. If there is a rational choice to be made between T and T', it "almost certainly" will have to be made on the basis of internal criteria. This "derives" the idealist posture. We also see that Einstein's methodology is not an ad hoc combination of convenient but inconsistent doctrines. The idealist and realist postures are part of one and the same response to the bankrupt methodology of late classical mechanics. Of course, these remarks do not show what inner perfection actually is. An Einsteinian need not be an opportunist, but he must remain, for the time being, a mystic.

Notice that the emergence of internal criteria (inner perfection) and the decline of external criteria (external confirmation) in rational choice comes about in a perfectly natural way when the objective content class gets small. This change of attitude may also be seen as a change of attitude with respect to the language $L_O$ with rigid semantic rules. For a positivist, $L_O$ is the only language capable of describing the real. This is its job. In Einsteinian methodology the $L_O$ language has a different function. Its job is to say what the theory is about, to define its intended models. The O terms serve only to <u>fix the reference</u> of all other terms in the language of the theory, to anchor them in the physical world. As science progresses, this gets done in an increasingly elegant way, with ever fewer rigid semantical rules.

APPENDIX I

1. The task in Appendix I is to express CPM as a regular theory in a formal, first order language, in conformity with the instructions given in the body of the paper. This will involve substantial departures from existing treatments.

   We may distinguish two broad approaches to the formalization of physical theories, the set-theoretic approach and the model theoretic approach (see [15]). The set-theoretical-predicate school constructs a set theoretical predicate which characterizes the mathematical structure of the theory. In the present case, a predicate "... is a CPM system" is constructed which holds of mathematical objects possessing the abstract structure of a CPM system. If the set theory is itself formalized in a first order language, then the elements in the extension of the predicate "... is a CPM system", are elements in some model for the set theory.

A two sorted model-theoretic approach is used, by Montague [13], in which the intended models consist of real numbers and particles. This approach has the advantage that the relation between mathematical and physical objects can be talked about in the formal theory. In the set theory approach the relation between the mathematical constructions and physical objects is either passed over entirely [12], or confined to the metalanguage [16].

2. A many sorted logic consists of (see [4]):

A. Logical symbols

1. Sentential connectives: $\Rightarrow$, $\Leftrightarrow$, $-$, $\vee$, $\wedge$.
2. Variables: for each sort $\delta$ there is a countable set of variables $v_1^\delta$, $v_2^\delta$,... of sort $\delta$.
3. Equality symbols: for each sort $\delta$, there will be the symbol $=^\delta$, said to be a predicate symbol of sort $\delta$.

B. Parameters

1. Quantifier symbols: for each sort $\delta$ there is a universal quantifier symbol $\forall_\delta$.
2. Predicate symbols: for each $n > 0$ and each n-tuple $(\delta_1,\ldots,\delta_n)$ of sorts, there is a set (possibly null) of n-place predicate symbols, each of which is said to be of sort $(\delta_1,\ldots,\delta_n)$.
3. Constant symbols: for each sort $\delta$ there is a set (possibly null) of constant symbols each of which is said to be of sort $\delta$.
4. Function symbols: for each $n > 0$ and each (n+1)-tuple $(\delta_1,\ldots,\delta_{n+1})$ of sorts, there is a set (possibly null) of n-place function symbols, each of which is said to be of sort $(\delta_1,\ldots,\delta_{n+1})$.

A <u>structure</u> for a many-sorted language, is a function $A$ on the set of parameters which assigns to each the appropriate object:

1. To the quantifier symbol $\forall_\delta$, $A$ assigns a nonempty set $|A|_\delta$, called the <u>universe</u> or <u>domain</u> of $A$ of sort $\delta$.
2. To each predicate symbol P of sort $(\delta_1,\ldots,\delta_n)$, $A$ assigns the relation $P^A \subset |A|_1 \times\ldots\times |A|_n$.
3. To each function symbol f of sort $(\delta_1,\ldots,\delta_{n+1})$, $A$ assigns a

function $f^A: |A|_1 \times \ldots \times |A|_n \to |A|_{n+1}$.

The definitions of satisfaction is straightforward and need not be reproduced here. Many-sorted logics are a mere notational variation of normal first order logics, and they may be reduced to the latter as follows: introduce a one-sorted language with all the parameters of the many-sorted language plus a one-place predicate symbol $Q_\delta$ for each sort. Translate all many-sorted sentences into one-sorted sentences by performing two changes; (1) replace each equality symbol $=_\delta$, with =, and (2) replace $\forall_\delta v_n (\text{—} v_n \text{—})$ with $\forall v (Q_\delta v \Rightarrow (\text{—} v \text{—}))$, where v is a variable whose substitution in this formula is permissible.

3. For the formalization of CPM we shall use a three-sorted language. Sort I will be the language of set theory. Axioms for this sort will be the usual axioms of ZF + choice. This part of the theory will not be explicitly presented, and we shall assume that the usual concepts of analysis; numbers, derivatives, integrals, etc. have been defined in this language. Whenever a formula involves a predicate or expression belonging to sort I, we shall feel free to designate this predicate or expression in the normal informal mathematical way.

The intended interpretations of $\forall_{II}$ and $\forall_{III}$ are space and time points respectively. We assume that we possess rigid semantical rules for identifying these point sets, and that these rules consist of coordinization procedures, mapping these point sets onto $\mathbb{R}^3$ and $\mathbb{R}^1$ respectively. It is not assumed that the coordinization has any natural or physical meaning. This lack of meaning is reflected in the freedom which we allow in metricizing these point sets. Whenever confusion cannot arise, sort subscripts will be omitted. To facilitate the exposition, we introduce the various predicates with a description of their mathematical form, instead of specifying this form by means of extra axioms.

O-vocabulary

The first three $L_O$ symbols provide us with the means of talking about spatial, temporal and massive points, respectively.

1. S(x) = r is a function of sort (II,I) taking spatial points into $\mathbb{R}^3$.

Ax. 1.1

S(x) is bijective.

2. T(y) = r is a function of sort (III,I), taking values in $\mathbb{R}^1$.

Ax. 2

T(y) is bijective.

It would be possible to introduce a separate sort for massive points, but this seems unnecessary. We can introduce massive points as functions from $|A|_{III}$ to $|A|_{II}$.

3. For $1 < i < n$, there is a set of functions $P_i(y) = x$ of sort (III,II).

Ax. $3.1_{ij}$

$$\exists_{II}x \; \exists_{III}y(P_i(y) = x = P_j(y)) \Rightarrow \forall_{II}z \; \forall_{III}w(P_i(z)=w \Leftrightarrow P_j(z)=w).$$

Axiom 3.1 says that two distinct massive points cannot occupy the same spatial position at the same time. The present approach to the definition of massive points may be traced back to Hertz [8]. Not all authors agree that univocity of position is essential to the notion of a massive point, and Suppes [12] for one, explicitly foreswears it. His attitude accords better with modern ideas of elementary particle behavior. On the other hand, axiom 3.1 is certainly consistent with the intended interpretation of CPM, and it simplifies the exposition at several points.

Hertz also recognized the possibility of introducing "unobserved masses". In effect, this means that additional functions of sort (III,II) representing unobserved masses should be introducible in the theoretical vocabulary. Although this is a sound proposal, it is not essential in an initial treatment.

In the presence of $\underline{S}(x)$ and $T(y)$, each $P_i(y)$ induces a function $\underline{P}_i$ from $\mathbb{R}^1$ to $\mathbb{R}^3$:

$$\underline{P}_i(r) = S(P_i(T^{-1}(r)))$$

This induced function can be used to require that the motions of the massive point be sufficiently smooth relative to the functions S(x) and T(y).

Ax. $3.2_i$

$\underline{P}_i(r)$ has continuous derivatives up to order $k = 3$.

It is convenient to possess the induced coordinate slot functions for describing the motion of the i-th massive point:

$$\underline{P}_i(r) = \langle p_i^1(r), p_i^2(r), p_i^3(r) \rangle$$

In this appendix we shall continue underlining induced concepts which might otherwise be taken to refer to $|A|_{II}$ or $|A|_{III}$, though in appendix II sections this distinction may be supressed.

TI-vocabulary

The TI-vocabulary is composed of the measure concepts and the kinematical concepts of velocity, acceleration, momentum and change of momentum.

The predicate for a time measure will be given in terms of a function similar to T:

4. $\psi(y)$ is a function of sort (III,I) taking values in $\mathbb{R}$.

Ax. 4.1

$\psi(y)$ is bijective.

For any measure we may define an induced function:

$$\underline{\psi}(r) = \psi(T^{-1}(r)).$$

This is just a bijection in the reals. T(y) induces the function:

$$\underline{T}(r) = T(T^{-1}(r)) = r.$$

We would like to require that the set of allowable time measures all report that time flows forward; however, we cannot require this without introducing an order relation in the domain $|A|_{III}$ in the O vocabulary c.f. [19]. We can require that all measures report forward flowing relative to the function T(y) by requiring that the induced functions of the measures form a monotone class, that is, that they are all strictly increasing functions. Increasing functions are differentiable except on at most a countable set, but as we are making stronger differentiability requirements elsewhere, we may as well add the further restriction to monotone functions which are differentiable. The desired property can be expressed as:

Ax. 4.2

$d\underline{\psi}/d\underline{T} = d\underline{\psi}/dr$ exists and is greater than zero.

In other words, the time rate as measured by $\underline{\psi}$ is greater than zero. Moreover, if $\underline{\psi}'$ is another conceivable time measure, then

$$d\underline{\psi}'/d\underline{\psi} = d\underline{\psi}'/dr \; dr/d\underline{\psi} > 0.$$

Switching from measure $\underline{\psi}$ to another measure $\underline{\psi}'$ which satisfies Ax. 4.2 can never produce a reversal or standstill in the flow of time.

The metrical predicate for space will be formalized as a metric tensor. To this end we introduce a 3 by 3 matrix of predicates:

5. $G_{ij}(x,y) = g_{ij}$ is a function from (II × III) to real numbers, for $1 \leq i, j \leq 3$.

Ax. $5_{ij}$

$g_{ij} = g_{ji}$,

$g_{ii} > 0$, Det $\begin{vmatrix} g_{11} & \cdots & g_{13} \\ \cdot & & \\ \cdot & \cdots & g_{33} \end{vmatrix} > 0$, for each $\langle x,y \rangle \in$ (II × III) and

and $G_{ij}$ is twice differentiable in all arguments.

Axiom 5 does not describe the essence of a tensor, for we have not specified a group of allowable coordinate transformations with respect to which the tensor is invariant. In effect, we have merely mentioned the coefficients $g_{ij}$ of a metric tensor in the cartesian coordinate system for $\mathbb{R}^3$. A full elaboration of the tensor concept is not essential here, and the reader is referred to Kreyszig [10] chapter V for details.

We shall regard the set of mass measures as derived from the set of positive real valued functions over massive points, but we cannot quite say it in this way because we cannot quantify over mass points as such.

6. $M(x,y) = r$ is a function from (II × III) to $\mathbb{R}_+$.

Ax. 6.1

$\forall_{III} y \, \forall_{II} x [ \bigvee_{1 \leq i \leq n} P_i(y) = x \Leftrightarrow M(x,y) > 0]$

It is also convenient to have an induced mass function for each massive point:

$\underline{M}_i(r) = M(S^{-1}(\underline{P}_i(r)), T^{-1}(r))$

Ax. $6.2_i$

$\underline{M}_i$ is differentiable.

Ax. 6.2 allows for the possibility that mass values vary with time.

The kinematical concepts depend upon a concept of arc length defined for any path in $\mathbb{R}^3$. Without losing generality, we shall introduce arc length predicates for each massive point.

7. $\underline{Arc_i}(q,r) = s$ is a function from $\mathbb{R}^1 \times \mathbb{R}^1 \rightarrow \mathbb{R}_+$.

Ax. $7_i$

$$\forall_I q,r,s \left[\underline{\mathrm{Arc}_i}(g,r) = s \Leftrightarrow s = \int_q^r \sqrt{\sum_{j,k=1}^{3} g_{jk}\, dp_i^j/d\underline{\psi}\; dp_i^k/d\underline{\psi}}\; d\underline{\psi}\right]$$

Henceforward we shall proceed as if $\underline{\psi} = \underline{T}$, and we shall write the more customary "t" for the generic time variable. If desired, the reader can recover formulas using $\underline{\psi}$ by referring to axiom 7. The expression "$\forall_I tA(t)$" will be understood to mean: "$\forall_I t(t \in \mathbb{R}^1 \Rightarrow A(t))$."

These definitions allow for a great deal of freedom in interpreting the metrical predicates. This accords with the fact that the coordination procedures are not assumed to have any physical meaning. However, we have not utilized all the freedom which is a priori available. One could let the functions $d\psi/dT$ depend on space and mass, and one could let the functions $g_{ij}$ depend on mass. Our reason for not doing this is to simplify the exposition.

8. $\underline{\mathrm{Vel}_i}(t) = q$ is a function $\mathbb{R}^1 \rightarrow \mathbb{R}^3$.

Ax. $8_i$

$$\forall_I t,\, q[\underline{\mathrm{Vel}_i}(t) = q \Leftrightarrow q = \langle dp_i^1/dt,\, dp_i^2/dt,\, dp_i^3/dt\rangle].$$

The proper treatment of acceleration requires some discussion. We cannot regard acceleration as a vector whose components are the time derivates of the components of the velocity vector, because we are no longer operating in Euclidean space. For a curve $(u^1(t), u^2(t), u^3(t))$ in Euclidean space, the velocity vector at $t_0$ is itself the derivate of the position vector: $\left.\frac{d\,u^i}{dt}\right|_{t=t_0}$. Now the position vectors whose limiting difference is the velocity vector are all attached to the same point, namely (0,0,0). In taking the derivative of the velocity, however, we are dealing with vectors attached to different points in space, e.g. the points along the motion of the massive point. In differential geometry, one says that these vectors belong to different tangent spaces. In order to look at the difference between the velocity vectors, say $v_1$ and $v_2$ at two instants we have to "carry" the vector $v_2$ into the tangent space of $v_1$, that is, we have to associate $v_2$ with some vector attached to the same point of space as $v_1$. In Euclidean space vectors are carried about so easily that one speaks of "free" or "unbound" vectors. Free vectors are determined by direction and magnitude, and not by the point of attachment. In Riemann space the notion of a bound vector is appropriate, and the rules for "carrying" a vector from one point in space along a given path to another point (the so-called Riemann connection) involve the metric tensor in an essential way. It is not generally the case that the vector in the tangent space of $v_1$ which results from "carrying" $v_2$ according to the Riemann connection has the same coordinate components as $v_2$. In short, the velocity vectors at the various points in time in a mass point's motion constitute a field of vectors which are spread out in space, and it is the time derivative of this field which must correspond to acceleration. The appropriate notion from differential geometry is the covariant derivative. If $\vec{a}(t)$ is a field of vectors along a curve C parametricized by t, the covariant derivative is defined:

$$\vec{a}(t) = \langle a^1(t),\ a^2(t),\ a^3(t)\rangle$$

$$(1)\quad D\vec{a}/dt = \langle Da^1/dt, Da^2/dt, Da^3/dt\rangle;\ Da^j/dt = $$

$$= \sum_{p=1}^{3} \sum_{k=1}^{3} da^j/dt + a^p\, \Gamma^j_{pk}\, du^k/dt.$$

The terms $\Gamma^j_{pk}$ are Christoffel symbols of the second kind and they are determined by the partial derivatives $\partial g_{ij}/\partial u^k$ of the metric tensor with respect to the spatial coordinates $u^k$; k = 1,2,3. In the above expression the terms $du^k/dt$ are the components of the tangents to C = $(u^1(t),\ u^2(t),\ u^3(t))$.

It is not appropriate here to further explicate the covariant derivative. Present purposes are served if its introduction is adequately motivated; the reader interested in more detail is referred to a standard source such as [10].

Acceleration will then be given in terms of the covariant derivative of the velocity vector:

9. $\underline{\text{Accel}_i}(t) = r$ is a function from $\mathbb{R}^1$ to $\mathbb{R}^3$.

Ax. $9_i$

$$\forall_I t, r\ \Big[\underline{\text{Accel}_i}(t) = r \Leftrightarrow r = \langle r^1, r^2, r^3\rangle;\ \text{and}$$

$$r^1 = d^2p_i^1/dt^2 + \sum_{j,k}^{3} \Gamma^1_{jk}\ dp_i^j/dt\ dp_i^k/dt;\ \text{and}$$

$$\vdots$$

$$r^3 = d^2p_i^3/dt^2 + \sum_{j,k}^{3} \Gamma^3_{jk}\ dp_i^j/dt\ dp_i^k/dt\Big].$$

The following axioms defining the predicates for momentum and the time derivative of momentum are self-explanatory.

10. $\underline{Mt}_i(t) = r$ is a function $\mathbb{R}^1 \to \mathbb{R}^3$.

Ax. $10_i$

$$\forall_I t, r \left[ \underline{Mt}_i(t) = r \Leftrightarrow r = \langle r^1, r^2, r^3 \rangle; \text{ and } r^1 = \underline{M}_i(t)\ d\underline{p}_i^1/dt \quad \vdots \quad \text{etc.} \right]$$

11. $\underline{\mathring{M}t}(t) = r$ is a function $\mathbb{R}^1 \to \mathbb{R}^3$.

Ax. $11_i$

$$\forall_I t, r\ [\underline{\mathring{M}t}_i(t) = r \Leftrightarrow r = d\underline{M}_i/dt\ \underline{Vel}_i(t) + \underline{M}_i(t)\ \text{Accel}_i(t)].$$

TN vocabulary

The vocabulary of non-interpreted theoretical terms contains only the predicate for force.

12. $F_{ij}(t) = r$ is a function from $\mathbb{R}^1$ to $\mathbb{R}^3$.

Ax. $12_{ij}$

$$\forall_I t, r\ [F_{ij}(t) = r \Rightarrow r \in \mathbb{R}^3].$$

Newton's second law, of which the first is a special case, is expressed in the following axiom.

Ax. $13_i$

$$\forall_I t \left[ \underline{\mathring{M}t}_i(t) = \sum_{j=1}^{n} F_{ij}(t) \right].$$

Following a suggestion of Montague we do not require $F_{ii}(t) = 0$. This allows for the representation of a net external force on the i-th massive point.

The notion of opposing forces in the sense of the third law must be handled with some delicacy, for again we have to compare two vectors which are bound at different points in space. Basically, one vector must be "carried over" to the tangent space of the other vector, where the angle between them may be determined by taking their inner product. However, the result of this carrying is not in general independent of the path over which the vector is transported. It is necessary to stipulate that the carrying take place along the shortest geodesic connecting the points at which the vectors are bound.

If G is a geodesic whose representation with respect to arc length s in the coordinate system $u^j$ is $u^j(s)$; $j = 1,2,3$; then we may consider the condition:

(2) $$da^j/ds + \sum_{m,i=1}^{3} a^m \Gamma^j_{mi} du^i/ds = 0.$$

This is a differential equation for the components $a^j$ of vectors defined along G and given as functions of s. By specifying coefficients $a^j(s_o)$ of a vector bound to the point of G corresponding to $s_o$, (2) determines coefficients for every other point of G. These solutions are said to be parallel in the sense of the Riemann connection with respect to G.

Two forces bound at given points with coefficients $a^j$ and $b^j$ may be called opposing if $a^j$ and $(-b)^j$, j = 1,2,3; are values of a solution to (2) when the unit tangent vectors $du^i/ds$ are derived from the shortest geodesic between the two points at which the vectors are bound. In this case we shall say that these vectors are opposing in the sense of (2).

<u>Ax. 14</u>$_{ij}$

$\forall_I t$ [$T_{ij}(t)$ and $F_{ji}(t)$ are opposing in the sense of (2)].

REMARK

Condition (2) defines opposing forces in a weak sense, for it requires only that opposing forces sum to the zero vector when they are transported into a common tangent space along the shortest geodesic between their points of attachment. In mechanics it is customary to require that the forces be directed along the line of centers as well. Under the weaker definition of opposing force it is not possible to prove that angular momentum is conserved. We have nevertheless adopted the weaker formulation, for in the widest sense of the term there are applications of classical mechanics in which opposing forces are not directed along the lines of centers, notably in electromagnetism (here, at least for point charges conservation of angular momentum is secured by attributing energy and momentum to the electromagnetic field).

This completes the formal exposition of CPM as a regular theory. The language of CPM is split into four sub-language:

(1) The logico-mathematical language $L_L$, $V_L$ = (all normal mathematical symbols definable in the language of set theory).

(2) The objective language $L_O$, $V_O = V_L + (S,T,P_i,\underline{P_i})$; $1 \leq i \leq n$.

(3) The theoretical-interpreted language $L_{TI}$, $V_{TI} = V_O +$ $(G_{jk}, ,M,\underline{M_i},\underline{Arc_i},\underline{Vel_i},\underline{Accel_i},\underline{Mt_i},\underline{Mt_i})$; $1 \leq j,k \leq 3$; $1 \leq i \leq n$.

(4) The theoretical-non-interpreted language, $L_{TN}$; $V_{TN} = V_{TI} + (F_{ij})$; $1 \leq i,j \leq n$.

Note that the induced predicates are not definable in the language $L_L$, even though their interpretations are within the domain $|A|_I$. This seems to express nicely the way in which mathematics applies to the physical world.

4. Let us recall the idea behind the O-vocabulary. According to definition 1 this is composed of predicates whose interpretation is complete and rigid. In the present case this entails that given any appropriate n-tuple of space points, time points, and numbers, we can decide once and for all without consulting the theory whether an O-predicate holds of this n-tuple. We regard this as unproblematical with respect to the O-vocabulary of section 3 (see however the remark to Ax 3.1).

Definitions 10 and 11 refer the question of objective and theoretical content to classes of conceivable intended models. We must examine the class of models which could conceivably be generated by the rigid $L_O$ rules. This class is not formally defined, for "conceivably" functions informally in this statement. Without delineating the notion of a conceivable application of the rigid semantical rules via axioms in $L_O$ it is impossible to do this in any meaningful way. This suggests that relative content is the only quantity which we can meaningfully calculate, where relativisation is taken with respect to the axioms in question. Of course, any particular relativisation could always be called into question if it did not adequately seem to capture the notion of a conceivable intended model

In spite of the fact that the answerable content questions involve only relative content questions, we hope to show that in a specific case, e.g. CPM, it is still possible to draw strong conclusions. With respect to the present exposition of CPM we declare that models in which $T_O$ (Ax. 1-3) or $T_{TI}$ (Ax. 1-11) do not hold are not very interesting. We consider therefore the objective content relative to $T_O$ and the theoretical content relative to $T_{TI}$.

Inspecting definition 13 where the notions of relative content are spelled out, the objective content question becomes: The objective content of CPM relative to $T_O$ is the class of conceivable intended models of $L_L + L_O$ which cannot be expanded to models of CPM minus those which cannot be expanded to $T_O$, that is, minus those which are not models of $T_O$ (since the $L_O$ vocabulary is already interpreted in these models). In other words the objective content class relative to $T_O$ is the class of models of $T_O$ which cannot be expanded to models of CPM.

Similarly the theoretical content of CPM relative to $T_{TI}$ is the class of conceivable intended models of $L_{TI}$ which cannot be expanded to models of CPM, minus those which cannot be expanded to models of $T_{TI}$, that is, minus those in which $T_{TI}$ do not hold (since, again, the $L_{TI}$ vocabulary is already interpreted). In other words, the theoretical content of CPM relative to $T_{TI}$ is the class of models of $T_{TI}$ (and this includes $T_O$) which cannot be expanded to models of CPM.

Note that this relativisation procedure will lead to counterintutive results if the theory is thrown into a logically equivalent form in which the axioms at the O or TI level are strengthened. For example, we might find an $L_O$ sentence which eliminates the theoretical vocabulary of CPM, that is, an $L_O$ sentence which holds in exactly those $L_O$ models which are expandable to CPM. If we consider the objective content of CPM relative to $T_O$ plus this sentence the result will of course be null.

We must now consider the class of conceivable applications of the rigid $L_O$ rules which lead to models of $T_O$, and we may interpret conceivability as mathematical conceivability. We consider the class of finite sets of class k = 3 functions $\mathbb{R}^1 \to \mathbb{R}^3$ which assign different points in $\mathbb{R}^3$ for every $t \in \mathbb{R}^1$ (as required by axiom 3). This class of mathematical objects contains all induced position functions $\underline{P}_i$ of all informally conceivable models of $T_O$, hence it is not too small.

On the other hand, if we took the class of finite sets of conceivable induced functions $\underline{P}_i$ as some proper subset of the above class, this would have the effect of introducing another axiom (if this subset is recursive) and this axiom should appear in the $T_O$ part of the theory. Until this is done the above proposal for translating the notion of an informally conceivable intended model is the only defensible one.

D14. The class of conceivable intended interpretations of $T_O$ (Ax. 1-3) is the class of finite sets of thrice differentiable functions $\underline{P}_1 \ldots \underline{P}_n : \mathbb{R}^1 \to \mathbb{R}^3$ such that

$$\{r \in \mathbb{R}^1 \mid \underline{P}_i(r) = \underline{P}_j(r),\ i \neq j,\ 1 \leq j, i \leq n\} = \emptyset.$$

D15. The theoretical content of CPM relative to $T_{TI}$ (Ax. 1-11) is the class of conceivable intended models of $T_{TI}$ which cannot be expanded to models of CPM.

D16. The objective content of CPM relative to $T_O$ (Ax. 1.3) is the class of conceivable intended models of $T_O$ which cannot be expanded to models of CPM.

REMARK:

This formalization is not entirely faithful to the program of formalizing CPM as a regular theory in at least the following respect. In separating the domains of space points and time points, we in fact lay claim to a semantically rigid, time independent method of identifying points of space. This is more than we can justify. A rigid semantic rule for interpreting "space-time point" is defensible on the horizon of late classical mechanics, but a rigid rule for identifying points of space in time is not.

Classical mechanics can be developed as a theory of an affine space-time manifold (see Trautman [18]). It is interesting to note that in this treatment force and (absolute) time are regarded as semantically rigid, and the affine connection is treated as theory-dependent! Such a treatment could be adopted to the present purposes by "theoretizing" time and force. I see no intrinsic difficulty in giving an adequate representation of CPM as a regular theory. Although the present treatment is simpler, it introduces an element of semantic rigidity, namely the "state of motion of the observer" which cannot be justified and which should be "theoretized" in a fully adequate treatment. I regard the present simplification as acceptable for the following reason: The interesting methodological consequences depend only on the objective content class becoming small. Additional theoretization will only make it smaller.

## APPENDIX II

The following results will be useful in the study of objective content.

THEOREM 1

A model of $T_{TI}$ composed of a single massive point with constant mass moving along a geodesic path with constant speed is a model of CPM when all forces are interpreted as the zero vector.

*Proof*. We require that $\dot{\vec{M}t} = \vec{0}$, and since $dM/dt = 0$, this entails that accel = 0. That is, for each coordinate function $p^i$, $i = 1,2,3$; we require that the differential equation (summation convention; sum over repeated indices):

$$d^2p^i/dt^2 + \Gamma^i_{lk}dp^l/dt\ dp^k/dt = 0$$

be satisfied. To express this in terms of arc length s (c.f. Ax.7), we may substitute:

$dp^i/dt = dP^i/ds \; ds/dt$

$d^2p^i/dt^2 = d^2p^i/ds^2 \; (ds/dt)^2 + dp^i/ds \; d^2s/dt^2.$

The requirement becomes:

$$(1) \quad (ds/dt)^2 \; [d^2p^i/ds^2 + dp^k/ds \; dp^q/ds \; \Gamma^i_{qk}] + dp^i/ds \; d^2s/dt^2 = 0$$

The bracketed expression is the differential equation for geodesics on a manifold characterized by the Christoffel symbols $\Gamma^i_{qk}$, when the function $p^i(s)$ describe a geodesic the bracketed expression equals zero. Similarly when the speed ds/dt is constant, $d^2s/dt^2 = 0$ and the requirement is satisfied.

THEOREM 2

If a single massive point is moving along a geodesic path, its acceleration is either zero or parallel to the velocity, i.e. in the direction of the motion.

*Proof.* The i-th coordinate of the acceleration is given by the LHS of equation (1). If the path in question is a geodesic, the bracketed expression is equal to zero, and we have:

$$\text{i-th coordinate of acceleration} = dp^i/ds \; d^2s/dt^2.$$

The vector with coordinate $(dp^i/ds)$ is just the unit vector tangent to the path. The acceleration vector is equal to the velocity vector multiplied by $(d^2s/dt^2)(ds/dt)^{-1}$.

The following result gives some insight into the objective content class for CPM. For I an open interval of real numbers, let us call a path segment $(P(t) \in R^3 : t \in I)$ simple if for $t_1, t_2 \in I$, $P(t_1) = P(t_2)$ implies that $t_1 = t_2$.

THEOREM 3

Let $(x^1, x^2, x^3)$ be cartesian coordinates in $\mathbb{R}^3$, let I be an open interval of $\mathbb{R}^1$, and suppose that P(t) is a simple path for $t \in I$, of class k = 3 (i.e. with derivatives up to order 3). If $dP/dt \neq 0$ for all $t \in I$, then there exist metric coefficients in $\mathbb{R}^3$; $g'_{ij}(x^1, x^2, x^3)$, such that P(t) describes a motion along a geodesic path with constant speed with respect to these coefficients.

*Proof.* There is a standard construction, applicable under the hypotheses of theorem 3, of (local) geodesic parallel coordinates, $u^1, u^2, u^3$.
With respect to these coordinates, P(t) may be written:

$$P(t) = (0,0,u^3(t)),$$

that is, the motion is along the $u^3$ coordinate, through the origin. Moreover, the Euclidean metric in $\mathbb{R}^3$ in the new coordinates takes the form:

$$ds^2 = (du^1)^2 + (du^2)^2 + g_{33}(u^1,u^2,u^3(du^3)^2.$$

(see [10], p. 145).

With respect to these coordinates we introduce new metric coefficients. Put $g'_{ij} = \delta_{ij}$, $i,j = 1,2$; and $g'_{3i} = 0$, $i = 1,2$. For $g'_{33}$ we stipulate:

$$g'_{33}((0,0,u^3)(t)) = 1/(du^3/dt)^2.$$

This defines $g'_{33}$ along the $u^3$ coordinate through the origin. It is meaningful because $du^3/dt = dP/dt \neq 0$; moreover, since $P(t)$ is simple, the dependence on $t$ can be eliminated. We extend $g'_{33}$ to the entire coordinate patch by requiring $\partial g'_{33}/\partial u^2 = \partial g'_{33}/\partial u^2 = 0$. For orthogonal coordinates $\Gamma^k_{jj} = \dfrac{2g_{jj}}{2g_{kk}\partial u^k}$ so that $\Gamma^2_{33} = \Gamma^1_{33} = 0$; and this in its turn entails that $P(t)$ describes a geodesic path ([10], p. 141). Moreover,

$$ds/dt = \sqrt{g'_{33}(du^3/dt)^2} = 1$$

The coordinates $u^i$ are local coordinates in $\mathbb{R}^3$. The proof is completed by extended these in any permissible way to all of $\mathbb{R}^3$.

A few heuristic considerations will suffice to give an idea of the content class for CPM relative to $T_O$. A model of $T_O$ with a single mass point whose velocity is never zero can be made a model of CPM by successive applications of theorem 3 to simple path segments of the mass point's trajectory. If there is more than one mass point, the restriction of velocity can be greatly relaxed by allowing interactions between the various particles.

Theorem 3 is important for two reasons. First, it shows that the (relative) content class of CPM cannot play a meaningful role in the rational evalution of CPM. Second, it suggests that this content class cannot be significantly increased by the addition of special force laws and constraints, as envisioned by the Sneed school. The problem with respect to objective content is not the generality of Newton's laws, the problem is the enormous freedom in the interpretation of the metrical predicates, especially the metric tensor $g_{ij}$ (the only metrical predicate used in these proofs). We do not need more axioms at the TN level, in order to generate objective content, rather, we need axioms at the TI level restricting the metric tensor. It is hard to imagine how such axioms could be justified, without restricting also the spatial coordination predicate S in the $L_O$ vocabulary.

NOTES

1) A great deal of effort has been directed toward giving a logical (syntactical or semantical) definition of theoreticity [16,9,14,17]. It is clear that the sort of theory-dependence exemplified by TI terms can only be defined pragmatically. Regarding TN terms, the following suggests itself. Let $\Delta$ be a class of possible worlds in which the non-theoretical vocabulary has been interpreted. Ideally the TI terms should be interpretable in every possible world $W \in \Delta$. TN terms, on the other hand, can only be "interpreted" by deriving their extension from the theory itself. This suggests that we could interpret TN terms only in those possible worlds which could be expanded to models of the theory, and this in turn suggests a semantic definition of TN theoreticity. However, this approach seems unacceptable to me. For a regular theory whose objective content is empty, the TI/TN distinction would collapse. Further, the extension of a TN term may be derivable from a proper part of the theory whose objective content is empty. For example, Mach and Hertz both define force explicitly via Newton's second law of motion, and the objective content of the second law by itself is empty. For this reason I prefer to define a semantic interpretation as a theory-external rule, i.e. a rule which does not consist of calculating values with the help of the theory. One simply says, then, that TN terms don't have semantic interpretations. This definition, however, is pragmatic (see [15]).

2) Consider two bodies, 1, and 2, in mutual interaction, with masses and accelerations respectively; $M_1, a_1, M_2, a_2$. If the bodies are isolated from interaction with any other body, then the third law of motion requires:

$$M_1 a_1 = -M_2 a_2.$$

Mach defines the mass ratio for these two bodies by the relation:

$$M_1/M_2 = -a_2/a_1$$

Let t'(t) be the readings of another clock which is running down relative to t, and suppose that dt'/dt exists for all t. Then $0 < dt'/dt < 1$. If $a_i = d^2x_i/dt^2$; then under the new clock we will observe the following accelerations and mass ratio:

$$a_i' = a_i \; dt/dt' + dx_i/dt \; d^2t/dt'^2;$$

(1) $M_1/M_2 = -a_2'/a_1'$

If $a_i' = 0$ for i = 2,1, the mass ratio (1) will be zero or undefined respectively. If the metric of space if fixed, and if we have a sufficient number of mutual interactions at our disposal, then by requiring that (1) be meaningful we require that $d^2t/dt'^2 = 0$. This means that we can write:

$$t' = bt + c \qquad b > 0.$$

3) If in note 2 the space metric is not given, then we cannot restrict the allowable clocks in this way to a linear transformation. Consider the transformation:

(2) $t' = t'(t)$; $dt'/dt > 0$ for all t

$$g'_{ij} = (dt'/dt)^2 \, \delta_{ij}; \quad \delta_{ij} = \begin{matrix} 0, \text{ if } i \neq j \\ 1, \text{ if } i = j \end{matrix}$$

We may regard the pair $\langle t', g'_{ij} \rangle$ as another possible choice of clock and meterstick (where the metrical coefficients $g'_{ij}$ are associated with the coordinates $x^j$). Then:

$$dx^i/dt' = (dx^i/dt)(dt/dt')$$
$$g'_{ii}(dx^i/dt')^2 = (dx^i/dt)^2.$$

This means that all velocities will remain unchanged under transformation (2). The clock t' and the meterstick $g'_{ij}$ induce apparent accelerations which cancel each other out. From the viewpoint of classical mechanics we must conclude that the metric for time is left totally undermined, and the metric for space is determined up to uniform expansion and contraction.

It is interesting to compare this with the situation in special relativity. There we have a space-time metric space, with distance element ds at our disposal. If:

$$(ds)^2 = (dx^1)^2 + (dx^2)^2 + (dx^3)^2 - (dt)^2,$$

then, since $(dt')^2 = (dt'/dt)^2(dt)^2$, the distance element resulting from the new clock and meterstick is:

$$(ds')^2 = g'_{11}(dx^1)^2 + g_{22}(dx^2)^2 + g_{33}(dx^3)^2 - (dt')^2$$
$$= (dt'/dt)^2 \, (ds)^2$$

This ambiguity is now expressed as a choice of a scale factor (not necessarily constant) for the space-time distance element.

4) A remarkable statement of this realistic attitude toward space can be found in the essay "Relativity and the Ether". We read there, for example:

"According to the general theory of relativity space is endowed with physical qualities; in this sense, therefore, an ether exists. Space without an ether is inconceivable. For in such a space there would not only be no propagation of light, but no possibility of the existence of scales and clocks, and therefore not spatio-temporal distances in the physical sense" [5 p. 204].

REFERENCES

1. Carnap, R. "Testability and Meaning" Philosophy of Science vol. 3 no. 4 (1936).
2. Cooke, R. "The Objective Content of Classical Particle Mechanics" Filosofische Reeks nr. 4 Universiteit van Amsterdam (1977)
3. Cooke, R. "Realism and Content" Kennis en Methode, jaargang 2, nr. 4,319-341
4. Enderton, Herbert B. A Mathematical Introduction to Logic, Academic Press New York (1972)
5. Einstein, A. The World as I See It, London (1941)
6. Ibid, "Autobiographical Notes" in Albert Einstein, Philosopher Scientist P.A. Schilpp (editor) New York (1951)
7. Hempel, C. "Inductive Inconsistencies" in Aspect of Scientific Explanation The Free Press (1951)
8. Hertz, Heinrich, The Principles of Mechanics Presented in a New Form New York 1956
9. Kamlah, Andreas "An Improved Definition of 'Theoretical in a Given Theory'" Erkenntnis 10 349-359 (1975)
10. Kreyszig, Erwin, Introduction to Differential Geometry and Riemann Geometry U. of Toronto Press (1968)
11. Mach, Ernst, The Science of Mechanics, La Salle Ill, (1960)
12. McKinsey, J.C.C., Sugar, A.C. and Suppes, P. "Axiomatic Foundattions of Classical Particle Mechanics" J. of Rational Mechanics and Analysis vol. 2 nr. 22, 253-272
13. Montague, Richard, "Deterministic Theories" in Decisions, Values and Groups, Washburn (editor), Permagon Press, 325-370 (1957).
14. Moulines, C. Ulises, "A Logical Reconstruction of Simple Equilibrium Thermodynamics" Erkenntnis 9 101-130 (1975)
15. Przelecki, Marian, "A Set Theoretic Versus a Model Theoretic Approach to the Logical Structure of Physical Theories" Studia Logika 33, 1 (1974)
16. Sneed, J.D. The Logical Structure of Mathematical Physics, Dordrecht (1971)

17. Stegmüller, W. Probleme und Resultate der Wissenschaftstheorie und Analytischen Philosophie, Band II, Theorie und Erfahrung; Zweiter Halbband, Theorien Strukturen und Theoriendynamik, Springer-Verlag, Berlin, (1973)
18. Trautman, A. "Foundations and Current Problems of General Relativity" in S. Deser and K. Ford (editors) Lecture on General Relativity, Prentice-Hall (1965)
19. Van Fraassen, B. Introduction to the Philosophy of Space and Time, Random House (1970)

# A COMPARISON OF TWO RECENT VIEWS ON THEORIES

E. Scheibe

Philosophisches Seminar der Universität
Nikolausberger Weg 9c
3400 Göttingen
Federal Republic of Germany

I.

In the following paper I am going to give a partial comparison of two recent proposals for the concept of a physical theory. The <u>first</u> proposal is due to Ludwig, and its original version appeared in Ludwig 1970 as a by-product of an attempt to give a physically satisfactory axiomatization of quantum mechanics. Meanwhile Ludwig has further developed his concept, and a self-contained presentation of it was given in Ludwig 1978. Using earlier approaches to the axiomatization of classical mechanics the <u>second</u> proposal was made by Sneed in connection with the so-called problem of theoretical terms (Sneed 1971). In contrast to the Ludwig approach which has remained isolated up to this very day, Sneed's conception has received much attention and was even presented as the way out of the difficulties that beset the orthodox view of scientific theories (see Stegmüller 1979 for an up-to-date account). In my own opinion this situation is nothing but an historical accident and does not in the least mirror the respective merits of the two approaches. Consequently, calling attention also to Ludwig's work certainly is one main purpose of my paper. But I have to admit from the outset that this purpose will be connected with a rather special interest in a foundational problem concerning the concepts in question. Although the solution of this problem will lead us to confront them, on account of the limited aspect chosen the resulting comparison will be very selective, and a thoroughgoing comparative appreciation, however welcome it would be, is beyond the scope of this paper.

The foundational problem I have in mind originates in a difference between the two programs according to which the corresponding concepts of a physical theory are to be reconstructed. According to Ludwig's program (the L-program) one of the things that have to be made explicit in the reconstruction of a theory is its language, and the way in which this has to be done is, as usual, formalization. On the other hand, according to Sneed's program (the S-program, where, for that matter, the 'S' may rather remind us of Suppes) the explication of a formal language as one of the elements of a theory is deliberately avoided. It is, of course, tempting to refer this situation to the logical empiricist's or - as it has been called - the received view of scientific theories. For it would then turn out that whereas the S-program was a deliberate move away from this view, the accordance of the L-program with the latter (which, by the way, ist not confined to the present poit) may rather have been a piece of preestablished harmony than a deliberate succession. But the details of these relations need not concern us here. What really matters is the fact that Ludwig's concept of a theory (the L-concept) is syntactical in the usual sense mentioned before whereas Sneed's concept (the S-concept) is not. Rather, as Sneed has put it, "the way of talking about scientific theories I am going to describe invites us to look at sets of 'models' for these theories rather than the linguistic entities employed to characterize these models" (Sneed 1976, p. 144, n 2). It is in this sense that his approach has been classified among the semantical approaches to the concept of a scientific theory (cf. Suppe 1974, p. 223, n 558).

At this point one would perhaps like to know the reasons that have been given in the S-program for abandoning the usual explication of the linguistic part of a theory. But I shall refrain from any direct comment on this matter. Although it will become clear in the course of the following consideration that these reasons cannot really be compelling this result will remain a side-issue of the paper. The main line of my argument will rather concern a problem that is caused by the difference between the L-concept and the S-concept as it was outlined a moment ago, namely the problem how the two concepts can be compared in view of this difference. This is the foundational problem to be solved in this paper, and although it certainly is not a very deep one and its solution will only be a first step towards a more complete comparison it will readily be admitted that something must be said about how to cope with the difference in question when undertaking a comparison of the L- and S-concepts. In fact the basic idea for a solution is simple enough and consists in just removing of perhaps rather in bridging the difference by either of two procedures of mutual adaptation: given the (syntactical) L-concept we may ask for a syntactical counterpart of the (semantical) S-concept and, vice versa, given the latter, one may look for a semantical counterpart of the former. Granted that these counterparts exist a common basis of comparison for the two concepts on the syntactical as well as on the semantical level will be prepared.

In order to realize this idea we shall have to recall 1) that theory elements typically distinguished by the L-concept are sentence forms (as physical axioms) in a formalized set theory, 2) that theory elements typically distinguished by the S-concept are classes of structures, e.g. the classes of all (physically) possible or (physically) possible partial models, and 3) that as far as the latter can be characterized by linguistic means at all they appear as the classes of all structures satisfying certain set theoretical sentence forms or - as the official wording runs - set theoretical predicates. Therefore, fixing the set theoretical basis for the L-concept (roughly) as the system of Zermelo-Fraenkel (ZF) it will be very promising to look for the syntactical counterpart of the S-concept on this basis. On the other hand, there are some difficulties in getting at the corresponding result on the semantical level. For one thing, in developing their S-concept the advocates of the S-program not only abjured the linguistic method in theory explication, but also did not take the trouble, indeed refused, to specify any formal framework for the presentation of their concept. As was evidenced by several misunderstandings this has obscured their attempt to a considerable degree, and, accordingly, a more formal account of the matter would be desirable. This, however, means that we have to look for a formal set theory comprehensive enough to have the classes of structures mentioned in 2) and 3) as its possible objects. Here a second difficulty comes up: As we shall see later on these classes are not sets in the sense of ZF. Therefore we have to look for a more comprehensive theory in which sets and genuine classes are distinguished. From the various extensions of ZF that are possible candidates to solve the problem the system of von Neumann and Bernays (VNB) will here be suggested as the basis for a precise formulation of the (semantical) S-concept and the semantical counterpart of the L-concept.

II.

The common basis of comparison for the syntactical version of the L- and S-concept is a suitable extension ZF' of ZF: In order to include all the mathematics that is used in physics (or at least in the physical theory under consideration) ZF is first extended by definitions of all the terms and predicates needed in that part of mathematics. Secondly, in order to make physical interpretation possible an infinite series of new constants $c_i$ for sets is added. On the basis of ZF' thus defined I shall now give a reformulation of the L-concept introducing some minor changes for its own sake and then suggest the syntactical counterpart of the S-concept, thereby making some adjustments for the sake of its comparison with the former. For both cases it should be borne in mind that no complete characterization of either concept is intended. In particular, to keep the presentation as simple as possible one very important feature of the L-concept, the uniform structures introduced in order to match the inaccuracies of measurement, will be omitted altogether (cf. Ludwig 1978, § 6).

To begin with, let T be a physical theory to be specified

according to the L-concept. In order to obtain what is usually called the axioms or - as Ludwig puts it - the mathematical theory of T (Ludwig 1978, §§ 2, 4 und 7) we first select a double series

$$(DL_1) \qquad X_1,\ldots,X_n,\ s_1,\ldots,s_m$$

from the additional constants $c_i$, abbreviating the former by the vector notation X and s respectively. They determine the primary language of T consisting of all structures from ZF' containing no other additional constants than the X and s. Secondly, a series of scale terms

$$(DL_1') \qquad \sigma_1(X),\ldots,\sigma_m(X)$$

(abbreviated by $\sigma(X)$) is chosen, i.e. terms constructed from their arguments (which, besides the X, may include defined terms of ZF') by successively applying one of the operations that yield a power set or a cartesian product. The first axiom of T is then given by

$$(L_1') \qquad s_1 \in \sigma_1(X),\ldots,s_m \in \sigma_m(X).$$

This typification renders the constants s structures of types σ over the basic constants X. The remarkable thing about the typification is that according to the choice of the terms $\sigma(X)$ it provides counterparts of all the predicates and terms of arbitrary arity and order as they appear in the various (even many-sorted) independent logical calculi. Axiom systems formulated in terms of these calculi as they are frequently used in presentations of the received view of theories are, therefore, easily translatable into the present frame work. Finally, a sentence

$$(L_1'') \qquad \alpha(X,s)$$

of the primary language is directly introduced as the second axiom, the axiom proper, of T. It has to fulfill the following condition of canonical invariance (which is automatically satisfied by the typification $(L_1')$): Defining the relation $iso_\sigma(X,s;X',s';f)$ to hold iff bijections $f^\sigma$, canonically (and obviously) determined by the bijections of the X onto the X', map the structures s of types σ over the X onto structures s' of the same type over the X', it is required that

$$(1) \qquad iso_\sigma(X,s;X',f) \dot{\Rightarrow} \alpha(X,s) \Leftrightarrow \alpha(X',s')$$

can be proved from ZF'. Combining $s \in \sigma(X)$ and $\alpha(X,s)$ in one sentence defined by

$$(2) \qquad \Sigma(X,s) \equiv s \in \sigma(X) \wedge \alpha(X,s)$$

we can sum up our requirements by saying that

$(L_1)$ $\Sigma(X,s)$

is admitted as an axiom of T iff it is a species of structures in the sense of Bourbaki (1968, Ch. IV). As regards the physical significance of the invariance property of the reader is referred to some relevant considerations in Scheibe 1981.

Turning now to a second part, the physically effective part, of T I am going to propose what seems to me a little improvement of the L-concept as it is presented by Ludwig. As will be seen in our third step certain empirical interpretation rules are provided for T. However, these rules will not in general give a physical meaning directly to the primitive constants X and s of the primary language but rather to certain terms dependent on them. Now, in a section about the physically effective part of a theory Ludwig describes the transition from a theory including such uninterpreted terms to another, physically equivalent theory in which all terms are interpreted: Here the interpretation rules are immediately applicable to the primitive constants of the primary language (Ludwig 1978, § 7.3). Therefore, the general situation will perhaps most adequately be described by explicitly introducing a secondary language of T from the outset. Let this language be determined by a new series

$(DL_2)$ $Y_1,\ldots,Y_k,\ t_1,\ldots,t_l$

from our additional constants. The main idea of connecting the Y and t with our primary language will be that of extending the primary species of structures $(L_1)$ by definitions. In order to extend the typification $(L_1')$ scale terms

$(DL_2')$ $\tau_1(X),\ldots,\tau_k(X),\ \rho_1(X),\ldots,\rho_l(X)$

are introduced and lead to the new typification

$(L_2')$ $Y \in \tau(X) \wedge t \in \rho(X)$

With the help of further terms

$(DL_2'')$ $P_1(X,s),\ldots,P_k(X,s),\ q_1(X,s),\ldots,q_l(X,s)$

entering into the definitions

$(L_2)$ $Y = P(X,s) \wedge t = q(X,s)$

the extension of the axiom proper $(L_1'')$ is obtained. Finally, if we require that the terms P and q are intrinsic with respect to $(L_1)$, $\tau(x)$ and $\rho(X)$ (cf. Bourbaki 1968, Ch. IV. 1.6) it can easily be shown that $(L_2)$ is canonically invariant and, consequently, the extension of $(L_1)$ by $(L_2)$ is again a species of structures. Moreover, it turns out that under this assumption $(L_2')$ is already a consequence of $(L_1)$ and $(L_2)$ in ZF'.

Up to this point the secondary language has been considered only in so far as it is connected with the primary language and as this connection leads to an extension of the primary species of structures $(L_1)$. We are now going to consider the secondary language also in its own right. As I mentioned before it is this language to which the empirical interpretation rules will directly be applied. This suggests to look at it as the empirical language of our theory T and to ask for empirical consequences of T in the sense of consequences of $(L_1)$ and $(L_2)$ in ZF' that are expressed in the secondary language. We may even ask whether there is such a thing as a strongest empirical consequence that as such is representative for the empirical content of T. It turns out that this question can very well be answered in the affirmative if we extend our previous data by a series of scale terms

$(DL_2''')$ $\quad \theta_1(Y),\ldots,\theta_1(Y)$

and require that the typification

$(L_2'')$ $\quad t \in \theta(Y)$

can be proved from $(L_1)$ and $(L_2)$ in ZF'. Given the terms $\theta$ we ask for a species of structures

$(L_{12})$ $\quad \Theta(Y,t) \equiv t \in \theta(Y) \wedge \beta(Y,t)$

in the secondary language which is the strongest consequence of $(L_1)$ and $(L_2)$ in the sense that

(A) $\quad \Sigma(X,s) \wedge Y = P(X,s) \wedge t = q(X,s) \vdash_{ZF'} \Theta(Y,t)$

(B) for all $\theta$-invariant $\gamma$:
if $\Sigma(X,s) \wedge Y = P(X,s) \wedge T = q(X,s) \vdash_{ZF'} \gamma(Y,t)$
then $\Theta(Y,t) \vdash_{ZF'} \gamma(Y,t)$

It can be proved that taking $\beta(Y,t)$ to be the sentence

$(L_{12}')$ $\quad \bigvee\xi\eta\ \Sigma(\xi,\eta) \wedge \bigvee f\ \ \mathrm{iso}_\theta(Y,t;P(\xi,\eta),\ q(\xi,\eta);f)$

$\Theta(Y,t)$ defined by $(L_{12})$ is a species of structures satisfying (A) and (B) and that any two $\Theta,\Theta_1$ satisfying (A) and (B) are equivalent in the sense that

(3) $\quad \Sigma(X,s) \wedge Y = P(X,s) \wedge t = q(X,s) \vdash_{ZF'} \Theta(Y,t) \Leftrightarrow \Theta_1(Y,t)$

Obviously, $(L_{12}')$ is very much like the Ramsey sentence of our axioms $(L_1)$ and $(L_2)$, the only difference being that, for reasons of invariance, the equalities in ($L_2$) are replaced by isomorphisms.

Concluding the sketch of the L-concept a third theory element has to be introduced consisting of the empirical interpretation rules (Ludwig: Abbildungsprinzipien, cf. 1978, § 5) that were already mentioned before. As usual these rules serve the purpose to connect the secondary (empirical) language of a theory T with results of observations, experiments or measurements that are obtained in a certain domain of application of T: Knowing the interpretation rules is equivalent to knowing how our experimental findings are to be written into the empirical language of our theory. Now, in general the procedure that is to be followed in getting at observational statements in the empirical language of T from the original meter readings may be very indirect in the sense that other physical theories, different from T, must be invoked as auxiliary theories (Ludwig: Vortheorien). But the only stage in this procedure that is wholly contained in T itself is the final output consisting of statements that can be made in the secondary language of T and that, accordingly, are considered to be directly given as far as T is concerned.

In order to give a formal account of these statements a third series

($DL_3$) $\quad a_1, \ldots, a_N$

of our additional constants $c_i$ must be distinguished as possible names for objects in the domain $\Delta$. Since we want our observational statements to pronounce relations between these objects it is required that

(4) $\quad \Theta(Y) = \mathrm{Pot}(\Theta^o(Y)),$

i.e. that the (Y) are at least of the first order over the Y. Consequently, the structures t , typified by ($L''_2$), are relations

(5) $\quad t_\lambda \subset \Theta^o_{\lambda 1}(Y) \times \ldots \times \Theta^o_{\lambda k_\lambda}(Y)$

where the (Y) are the components of a complete decomposition of (Y) as a cartesian product. Some but not all of the (Y) may be independent of the Y. An observational report (Ludwig: Abbildungsaxiome), i.e. a final output of applying the interpretation rules to certain experimental findings, is of the form

$$(L_3) \qquad \begin{array}{l} a_\nu \in \Theta^o_{\lambda i}(Y) \\ \langle \ldots, a_\nu, \ldots \rangle \in t_\lambda \text{ resp. } \notin t_\lambda \end{array}$$

In the first line objects $a_\nu$ are classified as belonging to the to the domain $\Delta$, and in the second line they are related by the structures $t_\lambda$ in genuine empirical statements. Our theory T is then provided with empirical content in the minimal sense that an extension of $(L_1)$ and $(L_2)$ or - equivalently - of $(L_{12})$ by $(L_3)$ may be inconsistent in ZF.

The brief outline of the (syntactical) L-concept of a physical theory given thus far will be sufficient for the present purpose. I come now to the syntactical (version of the) S-concept. It will be developed in as complete an analogy to the L-concept as is possible (cf. Sneed 1971, pp. 161 ff for the original presentation of the (semantical) S-concept). At the same time, to elucidate the connection with the semantical S-concept the standard notation used in more recent presentations of the latter (cf. Balzer and Sneed 1977) will be applied with the only difference that the usual symbols will be primed in order to remind the reader that they stand (not for semantical but) for syntactical entities.

In the first step again some properties of the axioms of T have to be specified. This will be somewhat more involved than it was in the case of the L-concept. For whereas $(L_1)$ is a statement about a single structure our new axioms will be statements about a set of structures of a given type.

Moreover, an additional feature is introduced into the new axiomatics by the so-called constraint of the S-concept. To begin with, let

$(DS_1)$ $\quad I_p'$

be one of the additional constants. $I_p'$ creates the primary language of T in the same sense as did the X and s. Furthermore, let (in vector notation as before)

$(DS_1')$ $\quad M_p'(\xi,\eta)$

be a typification of the $\eta$ with respect to the $\xi$. Thus $M_p'(\xi,\eta)$ is of the form $(L_1')$ but it will not be necessary to make this explicit.

We now want to dispose of two properties of sentences of the primary language that correspond to the typification and invariance characterizing a species of structures. With the general abbreviation

(6) $$\beta^*(Y) \equiv \wedge x,\ x \in y \Rightarrow \vee \xi\eta.\ x = \langle\xi,\eta\rangle \wedge \beta'(\xi,\eta)$$

the first property can be expressed by

(7) $$\gamma(I_p') \vdash_{ZF'} M_p^*(I_p'),$$

i.e. a sentence $\gamma(I_p')$ has this property iff it has the consequence in ZF' that I' is a set of structures of the type given by $M_p'$. The second property depends on a natural extension of isomorphisms of structures as they were introduced in connection with (1) to sets of structures of a given type: Let $ISO_{M_p'}(x,x_1;\{f_\lambda\}_\lambda)$ mean that x and $x_1$ are sets of structures with typification $M_p$', that the $f_\lambda$ constitute a family of isomorphisms between all structures of x and all those of $x_1$, and that they induce a bijection between the unions of all the base sets of structures in x and $x_1$ respectively. Then our new invariance condition is given by

(8) $$\mathrm{ISO}_{M_p'}(x,x_1;\{f_\lambda\}_\lambda) \vdash_{ZF'} \gamma(x) \Leftrightarrow \gamma(x_1)$$

With these definitions at our disposal we can now approach the two axioms of T. The first one is obtained with the help of a species of structures

($DS_1$") $$M'(\xi,\eta) \equiv M_p'(\xi,\eta) \wedge \alpha(\xi,\eta)$$

with $M_p$' as its typification. It reads

($S_1$') $$M^*(I_p')$$

and, as can easily be verified, the properties (7) and (8) are automatically satisfied for this axiom. It is otherwise with our second axiom

($S_1$") $$C'(I_p')$$

where C' is a unary formula of ZF' of which it is <u>required</u> that (7), (8) and, moreover,

(9) $$M_p'(\xi,\eta) \vdash_{ZF'} C'(\{<\xi,\eta>\})$$

are fulfilled. In this sense C' is called a (syntactical) <u>constraint for $M_p$'</u>. The conjunction ($S_1$) of our two axioms again has the properties (7) and (8). ($S_1$) exactly corresponds to the axiom ($L_1$) of T according to the L-concept.

Althought the systematic connection between the syntactical and the semantical version of the S-concept will be rigorously established later on it may be in order to stop for a moment and anticipate this connection in an informal way. As regards its first part our theory T is about a set $I_p$' of (physical) strucures. The semantical counterpart of $I_p$' does not appear in the original S-concept. Rather $I_p$' is here introduced for obvious reasons of analogy with the single structure X,s of the L-concept. In the terminology of the S-program $I_p$' would have to be called a set of intended theoretical (!) applications. As a consequence of ($S_1$'), viz. (7), $I_p$' is a subset of the class of (physically) possible models, this class being the extension of $M_p$'. By submitting $M_p$' to the condition of being a typification the concept of a class of possible models is slightly more general than the original and slightly more special than the modified concept of a theory-matrix or, for that matter, of a class of possible models according to the S-program (cf. Sneed 1976, p. 162; Balzer and Sneed 1977, p. 197). The extension of our M' is the class of all models in the sense of the S-concept, i.e. the subclass of all possible models satisfying the central law of T (here represented by $\alpha$ in ($DS_1$")). The requirement that M' be a species of structures imposes on $\alpha$ a condition of invariance that is not foreseen by the S-concept. Finally, the extension of the foregoing constraint C' is the constraint of the original S-concept. The invariance property (8) does not appear in this concept. In our approach it is conditioned by requiring M' to be a species of structures. On the other hand, recently new conditions have been imposed on C' to be a constraint for $M_p'$ (cf. Balzer and Sneed 1977, p. 196). Their (obvious) transcription into the syntactical framework may be left to the reader.

Taking now the second step in the development of the syntactical S-concept the analogy to the procedure that was followed for the L-concept suggests the introduction of a secondary language of T. It will be created by a new constant

($DS_2$) $\quad I_{pp}'$

corresponding to the Y and t of the L-concept. Just as the latter were defined by ($L_2$) through the X and s so $I_{pp}'$ will be defined through $I_p'$ by the definition

($S_2$) $\quad I_{pp}' = R'(I_p')$

where the term R' is now obtained in the following way:
We start with terms (in vector notation)

($DS_2$') $\quad \gamma_1'(\xi,\eta), \quad \gamma_2'(\xi,\eta)$

that are intrinsic with respect to the scale terms

($DS_2$") $\quad \rho_1(\xi), \quad \rho_2(\xi)$

and the species of structures M' and define R' to be the term

$$(10) \qquad R'(Y) = \{x \mid \bigvee \xi\eta, x = \langle\gamma_1'(\xi,\eta),\gamma_2'(\xi,\eta)\rangle \wedge \langle\xi,\eta\rangle \in Y\}.$$

As in the L-program the secondary language of the S-concept is considered to be the empirical language of T. And as before we may ask for empirical consequences of the axiom ($S_1$) that can be obtained in the empirical language if the definition ($S_2$) is added. To answer this question a new typification (corresponding to ($L_2$"))

($DS_2$''') $\quad M_{pp}'(\zeta_1,\zeta_2)$

is introduced. It is related to $M_p'$ by the requirement that

(11) $$M_p'(\xi,\eta) \wedge \zeta_1 = \gamma_1'(\xi,\eta) \wedge \zeta_2 = \gamma_2'(\xi,\eta) \vdash_{ZF'} M_{pp}'(\zeta_1,\zeta_2)$$

Given $M_{pp}'$ we now have to look for a sentence

($S_{12}'$) $\Lambda(I_{pp}')$

that, besides satisfying (7) and (8) with respect to $M_p'$, is the strongest consequence of T in the sense that

(a) $M^*(I_p') \wedge C'(I_p') \wedge I_{pp}' = R'(I_p') \vdash_{ZF'} A'(I_p')$

(b) for all $M_{pp}'$ - invariant $\gamma$:

if $M^*(I_p') \wedge C'(I_p') \wedge I_{pp}' = R'(I_p') \vdash_{ZF'} \gamma(I_{pp}')$

then $A'(I_{pp}') \vdash_{ZF'} \gamma(I_{pp}')$.

The solution of this problem is again a somewhat modified Ramsey sentence eliminating $I_p'$ in the premise of (a), namely

($S_{12}''$) $\vee y, M^*(y) \wedge C'(y) \wedge \vee\{f_\lambda\}_\lambda .\ ISO_{M_{pp}'}(I_{pp}', R'(y); \{f_\lambda\}_\lambda)$

and it is unique up to the equivalence

(12) $$M^*(I_p') \wedge C'(I_p') \wedge I_{pp}' = R'(I_p') \vdash_{ZF'} A'(I_{pp}') \Leftrightarrow A_1'(I_{pp}').$$

In the terminology of the S-program the set $I_{pp}'$ is the set of intended applications of T. On account of ($S_2$) and (11) $I_{pp}'$ is a subset of the extension of $M_{pp}'$, i.e. of the class of all possible partial models. ($S_{12}$), the strongest empirical consequence of our axioms, is a precise syntactical formulation of the so-called 'empirical claim' of T in the sense of the S-program. There is an obvious modification in so far as ($S_{12}''$) has been made invariant in the sense of (8) by replacing the equality of $I_{pp}'$ and $R'(I_p')$ by an isomorphism statement. The most decisive modification, however, that has been made in view of the original S-concept concerns the generality of the transition (11) from $M_p'$ to $M_{pp}'$ and, consequently, the generality of $M_p'$. In the S-program only the special case is considered where the $\gamma_i'$ are (normed) projections: They, as well as the $\rho_i$, are chosen to be

(13) $$r_1'(\xi,\eta) \equiv \xi, \quad r_2'(\xi,\eta) \equiv \eta'$$
$$\rho_1(\xi) \equiv Pot(\xi), \quad M_p'(\xi,\eta) \equiv \eta' \in \rho_2(\xi) \wedge \ldots$$

where, for some $m_1 \leq m$, $\eta'$ is $\eta_1,\ldots,\eta_{m1}$ if $\eta$ is $\eta_1,\ldots,\eta_{m1},\ldots,\eta_m$. Evidently, our generalization exactly correspondends to the situation as we met it in the L-program. It is partly conditioned by a different attitude towards theoretical quantities. But this is a matter that must be dealt with on another occasion.

Coming finally to the third step, the empirical interpretation of T, the first thing that has to be observed is that, whereas the foregoing development of the syntactical S-concept could almost immediately be read from the original concept once the general idea of a syntactical version was formed, the S-program does not contain any hints whatsoever as regards the formation of empirical interpretation rules. The reason is that the advocates of the S-program once they had abandoned the linguistic view of theories acquiesced in the idea that the theory elements they had introduced directly referred to physical objects and that no explicit interpretation of any language was necessary. Leaving it undiscussed whether this was a justifiable strategy the reintroduction of the linguistic aspect certainly reopens the question of interpretation. The general situation being the same as for the L-concept I shall confine myself to the concept of an observational report as it was formalized in ($L_3$) for the L-program. Using the material mode of speech and speaking very roughly we think of the elements of $I_{pp}'$ as being physical systems which in turn may be composed of objects such that for these objects our experimental findings can be expressed in observational statements. Accordingly we have to introduce additional constants

($DS_3$) $\quad Y_{KL}, t_{\lambda L}$ and $a_\nu$

and the observational reports assume the form

$\langle Y_{1L},\ldots,t_{1L}\rangle \in I'_{pp}$

($S_3$) the same as ($L_3$) for the $a_\nu$ with respect

to the $\langle Y_{1L},\ldots,t_{1L}\rangle$ and $M'_{pp}(Y_L,t_L)$

Thus the $t_L$ are typified with respect to the $Y_L$(vector notation!) by $M_{pp}'$, the latter satisfying (4), and then the $a_\nu$ are classified and submitted to empirical relations exactly as it was assumed in ($L_3$).

## III.

In the previous section only the first part of our main task has been fulfilled: Taking over the (original) L-concept of a theory with some minor changes and suggesting a syntactical version of the S-concept we have obtained a basis for comparing the two concepts on the syntactical level. We have now to tackle the second part and lay the foundations for comparison also on the semantical level. One way of doing this could be by this time to leave the (original) S-concept essentially untouched and to develop a semantical counterpart of the L-concept. However, as was already announced in the introduction I want to go beyond such a result: As in the previous section the S-program was violated by reintroducing the linguistic aspect into it so in this section I want to challenge the strategy of an informal presentation of the S-concept by giving a formal account of it.

This means that the following considerations will not be about any semantical entities in the usual sense. To call the S-program 'semantical' is a misnomer anyway (for which I do not want to charge its advocates). Strictly speaking a metatheory can only be called 'semantical' if it contains concepts typical for semantical relations. Since the S-program deliberately excludes a formal language from the theory elements that are to be made explicit the original S-concept cannot strictly be called 'semantical'. This concept is semantical only in the derivative sense that metalinguistic expressions used to enumerate the theory elements directly refer to the entities that would be the referents of an object language of the theory if such a language had been made explicit. Having done just this in the foregoing section our modified S-concept could indeed be rendered semantical in the strict sense by introducing a model of ZF' and trying to find a concept of interpretation according to which referents in the model are assigned to all the syntactical theory elements $M_p$',M'... etc. introduced in the previous section.

Apart from the fact that this program could not be realized with respect to ZF (see below) I shall refrain to enter the semantical domain altogether. Even the S-program had as one of its goals the clarification of the relations in which the various theory elements distinguished by that program stand to each other, and although this was actually done only in a naive way that eventually was called an 'informal axiomatics' (Stegmüller 1979, §§ 1 and 2) such an enterprise by its very nature is a formal or - for that matter - a syntactical one. Therefore it is wise not to mingle it with an aspect that, however valuable it may be in a different context, may easily lead to misunderstandings in matters of an essentially formal nature. The following example will perhaps be helpful in understanding the alternative that I am about to suggest.

Suppose that we were not concerned with the S-concept of theories but with the better known mathematical concept of groups. Then, by analogy, our enterprise would consists in 1) producing certain kinds of syntactical entities ...φ... (of ZF) defined by certain properties, 2) defining terms G(...φ...) and op (...φ...) depending on the foregoing entities, and 3) showing that if the ...φ... have their defining properties then the terms G(...φ...) and op(...φ...) satisfy the axioms for a group with G(...φ...) as its base set and op(...φ...) as its operation. The following is an example of such a procedure: 1') φ is a variable of ZF indicating a set; 2') G(φ) is the term for the set of all bijections of φ onto itself, and op(φ) is the term for the relation in which any three elements x,y and z of G(φ) stand iff z is the product (in the usual sense) of x and y; 3') is the proof that indeed a group has been obtained, namely the group of all transformations of φ. In general we would perhaps like to say that what is presented by 1) - 3) is a syntactical, more or less general procedure of constructing groups. Indeed, if we take the concept of groups as defined by the usual axioms as our starting point then the question 'are there any groups?' can be given a purely syntactical answer by pointing to the procedure 1) - 3) and give instances of it in the manner just illustrated.

Looking now at the S-concept of a theory in the light of the foregoing consideration it must be said that only part 1) of the construction procedure has been settled and some hints for 2) have been given in the previous section. The systematic exposition of 2) as well as that of 3) and, above all, the formalization of the S-concept that is presupposed by the whole procedure are still waiting for their presentation. However, as was already indicated in the introduction the execution of our program will not be possible without extending our formal framework. Although the S-concept, taken by itself, allows a formalization within ZF and can even be realized by syntactical models in the way that was outlined a moment ago the intended realizations cannot be obtained within ZF. For the intended syntactical realizations of the theory elements $M_p$,M,... etc. distinguished by the S-concept, i.e. the terms to be defined in part 2) of the construction procedure, are the formal extensions of corresponding predicates, viz. the $M_p$',M' ... etc. of section II. Now, as is well known, in order that the extension $\{x|Q(x)\}$ of a predicate Q exists a statement of comprehension

$$(14) \qquad \bigvee y \bigwedge x \, . \; x \in y \Leftrightarrow Q(r).$$

must be provable in ZF. But precisely this is not possible for the predicates entering our syntactical version of the S-concept. It has to be emphasized that this is not a shortcoming of our syntactical reconstruction but affects the original S-concept as regards its intended physical applications: There is not a single physical theory appearing among the physical examples given by the S-movement for which, say, the class $M_p$ of potential models or the class $M_{pp}$ of potential partial models is not a genuine class in the sense that for its defining predicate Q formula (14) can be disproved in ZF. (The same is, of course , true for the L-concept for which, however, no 'semantical' version has been claimed to exist.)

We therefore have to look for a formal framework that allows for the distinction between sets and genuine classes. Among the various extensions of ZF that have been suggested in this respect I choose the theory of von Neumann and Bernays (VNB) for the simple reason that it includes the kind of classes we want to be included and essentially nothing beyond. The character of a class extension of ZF can perhaps most suitably be asserted by accepting a two-sorted version of VNB (cf. Fraenkel et al. 1973, pp. 119 ff): As before small letters will be used as variables for sets, and from now on the capitals A,B,... will be variables indicating classes. Under this assumption the axioms (and theorems) of ZF can be taken over without modification. The most important axiom characteristic of VNB is the axiom of prediative comprehension for classes. More precisely this is the axiom schema

(14') $\bigvee A \bigwedge x \,.\, x \in A \Leftrightarrow Q(x)$.

where $Q(x)$ is any formula which does neither contain A nor any other bound class variables. It is obvious that this axiom of comprehension goes far beyond the possibilities of ZF: Whereas for many formulas $Q(x)$ there may be no set containing precisely the elements x for which $Q(x)$ our new axiom admits a class $\{x|Q(x)\}$ for every Q of the kind described.

In section II we have seen that species of structures in ZF play a distinguished role in the presentation of the syntactical versions of our two concepts of theories. It turns out that a generalized concept of species of structures, adapted to our new framework VNB, can be favourably employed also for the formalization of the semantical concepts. Let us therefore briefly look for a natural generalization. As before we conceive of a species of class structures - as it may now be called - as being a formula

(15) $F_{typ}(\varphi,\psi) \wedge F(\varphi,\psi)$

where $F_{typ}$ is the typification, F the axiom proper and $\varphi$ and $\psi$ are vectors for class or set terms. The principal modification that is forced upon us in view of the new situation in VNB concerns the typification: There is no problem in forming cartesian products of classes with the help of (14'), the product of A and B being just the class of pairs $\langle x,y\rangle$ with $x \in A$ and $y \in B$. But the power class Pot(A) of a class A cannot be the class of its subclasses (which would violate (14')) but only the class of the subsets of A. Defining scale terms as before with respect to the new concepts of cartesian product and power class we have still to generalize the typification $\psi \in \sigma(\varphi)$ itself since this formula would the $\psi$ restrict to be sets. Although this is not to be excluded genuine classes must be allowed as typified classes and this is achieved by allowing the typification also to be of the form $\psi \subset \sigma(\varphi)$. As regards the axiom proper F it is easily seen that the invariance condition connected with (1) can be taken over almost verbally. We shall presently come to see that there may be some reasons to drop the invariance condition as part of the concept of a species of class structures if it is used in the definition of our two concepts of a physical theory.

Having laid the new foundations we can again attend to these concepts and first the L-concept. According to our preparations what was called the semantical version of the L-concept will now be defined by an axiom system in VNB or rather - similar to the situation in section II - in a suitable extension by definitions VNB' of VNB. As already indicated the axiom system can be given the form of a species of structures (15) where the arguments will be taken to be new class or set constants added to VNB: There will be two basic class constants $\Sigma_p^o$ and $\Sigma_{pp}^o$, three typified class constants $\Sigma^o$, U and $\Theta^o$ as well as two typified set constants $x^o$ and $y^o$ with the typifications

$$\text{(16)}\qquad \begin{array}{l} \Sigma^o \subset \Sigma_p^o,\quad U \subset \Sigma_p^o \times \Sigma_{pp}^o,\quad \Theta^o \subset \Sigma_{pp}^o \\ x^o \in \Sigma_p^o,\quad y \in \Sigma_{pp}^o \end{array}$$

The axioms proper are given by

$$\text{(17)}\qquad U\colon \Sigma^o \to \Theta^o,\ x^o \in \Sigma^o,\ \langle x^o, \Sigma^o\rangle \in U$$

where the first member means that U is a mapping from $\Sigma^o$ into $\Theta^o$. Up to this point we would have an invariant species of class structures and a very simple one at that. Since $\Sigma_p^o$ and $\Sigma_{pp}^o$ are meant to be the extensions of the typifications ($L_1$') and ($L_2$") we could go on in our list of axioms by requiring

$$\text{(17')}\qquad \Sigma_p^o \subset S^n \times S^m,\quad \Sigma'_{pp} \subset S^k \times S^l$$

where S is the class of all sets. These axioms, however, would no longer be invariant. At the same time, they would make the semantical L-concept dependent on additional parameters (here: numbers). Whether this is a desirable consequence remains to be seen.

The L-concept developed in the previous section can now be shown to be a syntactical model of the axiom system (16) and (17) in the sense of the procedure described at the beginning of this section. Using the definition schema

$$\text{(18)}\qquad Q^o = \{x \mid Q(x)\}$$

the constants $\Sigma_p^o$, $\Sigma_{pp}^o$; $\Sigma^o$, U, $\Theta^o$, $x^o$ and $y^o$ are defined by substituting for Q(x)

$$\text{(19)}\qquad \begin{array}{l} \bigvee\xi\eta\ .\ x = \langle\xi,\eta\rangle \wedge \eta \in \sigma(\xi). \\ \bigvee\varphi\psi\ .\ x = \langle\varphi,\psi\rangle \wedge \psi \in \theta(\varphi). \\ \bigvee\xi\eta\ .\ x = \langle\xi,\eta\rangle \wedge \Sigma(\xi\ \eta). \\ \bigvee\xi\eta\varphi\psi\ .\ x = \langle\langle\xi,\eta\rangle,\langle\varphi,\psi\rangle\rangle \wedge \varphi = P(\xi,\eta) \wedge \psi = q(\xi,\eta) \\ \qquad \wedge\ \eta \in \sigma(\xi) \wedge \psi \in \theta(\varphi). \\ \bigvee\varphi\psi\ .\ x = \langle\varphi,\psi\rangle \wedge \theta(\varphi,\psi). \\ x = \langle X,s\rangle \\ x = \langle Y,t\rangle \end{array}$$

in that order. Here the last two definitions are adapted to the schema (18) for reasons of uniformity and could instead be given directly in the simpler form $x^o = \langle X,s\rangle$ and $y^o = \langle Y,t\rangle$. It is now very easy to show that (16) and (17) are consequences of these definitions and the assumptions made in section II about the syntactical entities entering the definiens of (18). On the other side, it is also obvious that not all of these assumptions are actually needed to obtain (16) and (17), e.g. the assumption (B) for $\Theta$ is not. A consequence that would correspond to it would seem to say that among the classes into which $\Sigma^o$ is mapped by U the class $\Theta^o$ in some sense is the smallest. But it does not seem possible to define this sense without reference to the type $\theta$ appearing in $\Theta$ which again would make the L-concept itself dependent on external parameters.

As regards the S-concepts we first develop the precise semantical analogue of our syntactical S-concept of section II. Since we have taken the liberty of a few modifications it is to be expected that the result will deviate from the original S-concept in some respects. The species of structures representing the desired axiomatics has two basic class constants $M_p$ and $M_{pp}$ as well as the typified class constants M,C,r,A and the set constants $I_p$ and $I_{pp}$ with the typifications

$$M \subset M_p,\ C \subset pot(M_p),\ r \subseteq M_p \times M_{pp},\ A \subset pot(M_{pp}) \tag{20}$$
$$I_p \in Pot(M_p),\ I_{pp} \in Pot(M_p)$$

Introducing the term

$$R(y) \equiv \{x_1 \mid \vee x.x \in y \wedge \langle x,x_1\rangle \in r.\} \tag{21}$$

the axioms proper read

$$\begin{array}{l} \wedge x.x \in M_p \rightarrow \{x\} \in C \\ r: M_p \rightarrow M_{pp} \\ \wedge y.y \in pot(M) \cap C \rightarrow R(y) \in A. \\ I_p \in pot(M) \cap C \\ I_{pp} \in R(I_p) \end{array} \tag{22}$$

Bringing into play the syntactical data of section II and using again the definition schema (18) a syntactical model of the species of class structures (20)-(22) is obtained if the constants $M_p$, $M_{pp}$, M,C,r,A,$I_p$ and $I_{pp}$ are defined by substituting for Q(x)

$V\xi\eta.\ \ x = \langle\xi\ \eta\rangle \wedge M_p{}'(\xi,\eta).$

$V\zeta_1\zeta_2.\ \ x = \langle\zeta_1,\zeta_2\rangle \wedge M_{pp}{}'(\zeta_1,\zeta_2).$

$V\xi\eta.\ \ x = \langle\xi_1,\xi_2\rangle \wedge M_{pp}{}'(\xi_1,\xi_2).$

$V\xi\eta.\ \ x = \langle\xi,\eta\rangle \wedge M'(\xi,\eta).$

$C'(x)$

(23) $V\xi\eta\zeta_1\zeta_2\ .\ \ x = \langle\langle\xi,\eta\rangle\ \langle\zeta_1,\zeta_2\rangle\rangle \wedge \zeta_1 = r_1'(\xi,\eta) \wedge \zeta_2 = r_2'(\xi,\eta)$
$\wedge\ M_p{}'(\xi,\eta) \wedge M_{pp}'(\zeta_1,\zeta_2).$

$A'(x)$

$x \in I_p{}'$

$x \in I_{pp}{}'$

in that order, Relating (21) to (10) in an obvious way it is again easy to verify that (20) and (22) are consequences of these definitions and the asspumptions about the syntactical S-concept.

Let us finally review the essential modifications that we have made with respect to the original S-concept. There is first the generalization of the mapping r: In the original S-concept r is a projection and as such it is a mapping from $M_p$ onto $M_{pp}$. The latter property could, of course, be required in (22) and it could be proved from the syntactical concept if we would require the reversal of (11) with an existential quantification over ξ and η. Whether this restriction would be acceptable also in the general case that we have considered remains to be seen. Secondly, the invariance conditions of our syntactical S-concept are alien to the S-concept. If they are dropped in ($S_{12}$) and, consequently, the ISO-formula in ($S_{12}$") is replaced by an equality then the third axiom of (22) can be replaced by the equality

(22a) $A = \{x \mid \forall y.\ \ y \in \mathrm{Pot}(M) \cap C \wedge x = R(y)\}$

which, obviously, is stronger than the former. In proving (22a) from the modified ($S_{12}$) the condition (b) must be used. Thirdly, our $I_p$ does not occur as a theory element according to the original S-concept. In dropping it the last axiom of (22) has to be replaced by

(22b) $I_{pp} \in A$

Furthermore, we have omitted axioms corresponding to (17') of the L-concept and expressing the matrix character of $M_p$ and $M_{pp}$. They could be included only at the expense of the invariance property of the species of class structures representing the S-concept.

Apart from these modifications of a rather technical nature there was, finally, the basic methodological distinction between a syntactical and a semantical version of the S-concept and the formalization of the latter. Insofar as this, too, is a deviation from the original S-program it was dictated by the desire to have a basis of comparison of the S-concept with the L-concept of a physical theory. If it will have the side effect to provoke a stricter articulation of the S-program on the part of its advocates, so much the better. Even the comparison with the L-program for which the foundations have been laid in this paper but which remains to be done might be a useful contribution to this end.

## LITERATURE

Balzer, W., and Sneed, J.D.: Generalized Net Structures of Empirical Theories I. Studia Logica 36 (1977) 195-211

Bourbaki, N.: Theory of Sets. Paris 1968

Fraenkel, A.A., Bar-Hillel, Y., and A. Levy: Foundations of Set Theory. Amsterdam [2]1973

Ludwig, G.: Deutung des Begriffs "physikalische Theorie" und axiomatische Grundlegung der Hilbertraumstruktur der Quantenmechanik durch Hauptsätze des Messens. Berlin 1970

Ludwig, G.: Die Grundstrukturen einer physikalischen Theorie. Berlin 1978

Scheibe, E.: Invariance and Covariance. To appear in: Festschrift for Mario Bunge. Ed. J. Agassi and R. Cohen

Sneed, J.D.: The Logical Structure of Mathematical Physics. Dordrecht 1971

Sneed, J.D.: Philosophical Problems in the Empirical Science of Science: A Formal Approach. Erkenntnis 10 (1976) 115-46

Stegmüller, W.: The Structuralist View of Theories. Berlin 1979

Suppe, F.: The Search for Philosophic Understanding of Scientific Theories. In: The Structure of Scientific Theories. ed. F. Suppe. Urbana, Ill., 1974. 3-241

WHAT DO WE KNOW FROM LIGHT-EXPERIMENTS ABOUT THE PRINCIPLE OF RELATIVITY AND THE LIGHT-PRINCIPLE?

J. Pfarr

Institut für Theoretische Physik
der Universität zu Köln
5000 Köln 41, Zülpicher Str. 77
Federal Republic of Germany

## I. INTRODUCTION

Many derivations of the Lorentz-transformation are known in the literature, and in most of these analyses the basic invariance property is that the speed of light in inertial frames is independent of the motion of the source (PCL). The second set of invariances is governed by the Principle of Relativity (RP) saying that inertial frames cannot be distinguished from each other by internal processes and measurements[1)]. The validity of these two principles reduces the degrees of freedom for the transformation between Cartesian coordinate systems of different inertial frames; detailed analysis uniquely leads to the Lorentz-transformation.

In 1949 H.P. Robertson tried to replace several postulates of the two principles by properties gained from empirical experiences[2)]. In his deduction of the Lorentz-transformation the remaining postulates and assumptions, in comparison with the complete form of the principle, are just weak enough to allow a three-parameter family of transformations. And these parameters are uniquely determined by means of the results of three types of experiments: the Michelson-Morley (MM)[1887], the Kennedy-Thorndike (KT) [1932] and the Ives-Stilwell (IS) [1938] experiments. All of these experiments allow us to determine relevant parameters of the transformation to second order in v/c.

Robertson discussed the results of the three experiments - as they were historically performed - in <u>one</u> inertial frame (represented by the earth). He disregarded that the experiments can in principle be performed in <u>arbitrary</u> inertial frames. It will be shown in this paper that under much weaker assumptions than those given by Robertson the results of the three experiments - performed in each of two inertial frames in relative motion - will lead to the Lorentz-transformation (to second order in v/c). Under the complete renunciation of both the RP and the PCL we will show that the results of these three light-experiments can by themselves give evidence to the validity of the RP as well as the PCL.

## II. THE TRANSFORMATION BETWEEN INERTIAL FRAMES

### 1) Coordinate Systems within Inertial Frames

We start the following considerations with the notion of inertial frames. An inertial frame is defined by the following properties:

- it is a rigid frame of reference[3], which at any place in space is equipped with measuring rods and clocks to measure lenghts and local times[4].
- physical space is homogeneous and isotropic, and the geometry as measured by means of the measuring rods is assumed to be the Euclidean geometry.
- uninfluenced neutral test particles move along straight lines of this geometry.

Within a given frame of reference I, we can introduce spatial coordinate systems by mapping any triple of coordinates $x^k$ (k = 1,2,3) to some point in space. Together with the times $t_{(x^k)}$ defined by the local clocks we can thus define a spatio-temporal coordinate system $K(x^k, t_{(x^k)})$ in I. We will assume that the spatial coordinates do not depend on time themselves, i.e. we will deal with time-independent or comoving coordinate systems.

The system of coordiantes $K(x^k, t_{(x^k)})$ in I is not unique. By means of the transformation

$$T(f) \qquad \begin{aligned} x^k &\to \tilde{x}^k = f^k(x^j, t) \\ t &\to \tilde{t} = f^o(x^j, t) \end{aligned} \qquad \text{(II.1)}$$

we can introduce new coordinates with arbitrary functions $f^k, f^o$. If the new system $\tilde{K}(\tilde{x}^k, \tilde{t}_{(\tilde{x}^k)})$ is supposed to be comoving as well, then the new spatial coordinates $\tilde{x}^k$ must not depend on t. In this case (II.1) reads

$$\begin{aligned} x^k &\to \tilde{x}^k = f^k(x^j) \\ t &\to \tilde{t} = f^o(x^j, t). \end{aligned} \qquad \text{(II.2)}$$

Let the time coordinate in $K(x^k, t)$ in the frame I be chosen such that the straight-line motion of the uninfluenced test bodies is uniform with respect to t, i.e. the particles move with constant velocity with respect to t[5]. It can be found empirically that all motions of uninfluenced test particles are uniform with respect to each other. Therefore a time coordinate t can be introduced in I globally: $t := t_{(x^k)}$.

In addition to the condition II.2 the transformation T(f) will now be chosen such that the inertial character of the trajectories for the moving test bodies will be invariant: Motions along straight lines which are uniform with respect to t in $K(x^k,t)$ have to be uniform with respect to $\tilde{t}$ in $\tilde{K}(\tilde{x}^k,\tilde{t})$, as well. It can be shown that this requirement reduces the admissible transformations (II.2) to linear transformations between K and $\tilde{K}$[6].

$$x^k \rightarrow \tilde{x}^k = f^k{}_1 x^1$$
$$t \rightarrow \tilde{t} = \frac{1}{c} f^o{}_1 \cdot x^1 + f^o{}_o t \qquad \text{(II.3)}$$

$f^k{}_1$ are arbitrary parameters with $f^o{}_o \neq 0$, not all $f^k{}_1 = 0$.[7]

The following considerations are based upon experiments whose performance requires at most two spatial directions which are orthogonal to each other. Therefore the transformation T(f) can be restricted to two spatial dimensions, and under renunciation of shears, rotations and spatial reflections we get a five parameter transformation

$$x \rightarrow \tilde{x} = f^1{}_1 x$$
$$y \rightarrow \tilde{y} = f^2{}_2 y \qquad \text{(II.4)}$$
$$t \rightarrow \tilde{t} = f^o{}_o t + \frac{1}{c} f^o{}_1 x + \frac{1}{c} f^o_2 \; y$$

2) Coordinate Transformations between Different Inertial Frames

Let I and I' be two inertial frames, $K(x^j,t)$ and $\tilde{K}'(\tilde{x}^{j'},\tilde{t}')$ two corresponding inertial coordinate systems as defined in the preceding section. Let the spatial orientation of the coordinate systems be chosen such that the relative motion is along the respective x-axes, and let $v := dx/dt \,|_{d\tilde{x}^{k'}=0}$ be the velocity of I' with respect to I.

The inertial character of the frames of reference I and I' as well as the isotropy of space and the homogeneity of space and time in each frame reduce the transformation between K and $\tilde{K}'$ to the form

$$\tilde{x}' = k(v)(x-vt)$$
$$\tilde{y}' = \lambda(v) y \qquad \text{(II.5)}$$
$$\tilde{z}' = \lambda(v) z$$
$$\tilde{t}' = \mu(v)(t-\alpha(v)\nu x - \beta(v) vy - \gamma(v) vz)\text{[8]}.$$

Here we have adopted the convention that the origins of the coordinate systems coincide. In addition, we shall require that for v = 0 (II.5) reduces to the identity.

$k, \lambda, \mu, \alpha, \beta, \gamma$ are arbitrary symmetric functions of v, of which $k, \mu$, and $\lambda$ satisfy the initial condition $k(0) = \mu(0) = \lambda(0) = 1$. For the same reasons as in section (1) we restrict ourselves to two spatial directions:

$$T_{II'} \quad \begin{aligned} \tilde{x}' &= k(v)(x-vt) \\ \tilde{y}' &= \lambda(v) y \\ \tilde{t}' &= \mu(v)(t-\alpha(v)vx-\beta(v)vy). \end{aligned} \qquad \text{(II.6)}$$

II.6 is again a five parameter family of transformations. However, while T(f) represents the change of the coordinate system within a given inertial frame I, $T_{II'}$ denotes the transformation between two coordinate systems which belong to different inertial frames.

The parameters of $T_{II'}$ will be determined by the results of particular experiments. For purely technical reasons and reasons of clarity it is useful to use a set of parameters $f^1{}_2, f^2{}_2, f^o{}_o, f^o{}_1, f^o{}_2$ instead of the set $k, \mu, \lambda, \alpha, \beta$ above.

The two sets of parameters are connected by the following transformation

$$k(v) = f^1_1(v)/\sqrt{1-v^2/c^2}, \quad \alpha(v) = \frac{1-\frac{f^o_1}{f^o_o}\frac{c}{v}}{c^2\left(1-\frac{f^o_1}{f^o_o}\frac{v}{c}\right)}$$

$$\lambda(v) = f^2_2(v), \quad \beta(v) = \frac{f^o_2\sqrt{1-v^1/c^2}}{cv\, f^o_o\left(1-\frac{f^o_1}{f^o_o}\frac{v}{c}\right)} \qquad \text{(II.7)}$$

$$\mu(v) = f^o_o(v)\left(1-\frac{f^o_1}{f^o_o}\frac{v}{c}\right)/\sqrt{1-v^2/c^2}$$

The inverse transformation reads:[9)]

$$f^o_o = \mu(v)(1-\alpha(v)v^2)/\sqrt{1-v^2/c^2} \qquad f^o_1 = \mu(v)\frac{v}{c}(1-\alpha(v)c^2)/\sqrt{1-v^2/c^2}$$

$$f^1_1 = k(v)\sqrt{1-v^2/c^2} \qquad f^o_2 = -\mu(v)\beta(v)vc \qquad \text{(II.8)}$$

$$f^2_2 = \lambda(v).$$

$f^1{}_1, f^2{}_2$ and $f^o{}_o$ are symmetric, $f^o{}_i$ antisymmetric functions of v.

If in I' a coordinate system K'(x',y',t') is hypothetically introduced which is defined by

$$T_L \quad \begin{aligned} x' &= (x-vt)/\sqrt{1-v^2/c^2} \\ y' &= y \\ t' &= \left(t-\frac{vx}{c^2}\right)/\sqrt{1-v^2/c^2} \end{aligned} \qquad \text{(II.9)}$$

then II.6 reads

$$\begin{aligned} \tilde{x}' &= f'_{,}x' \\ \tilde{y}' &= f^2{}_2 y' \\ \tilde{t}' &= f^o{}_o t' + \frac{f^o{}_1}{c} x' + \frac{f^o{}_2}{c} y'. \end{aligned} \qquad \text{(II.10)}$$

Obviously $T_L$ represents the Lorentz-transformation. However, as long as the parameters f in II.10 are not fixed to $f^i{}_i = 1$ and $f^o{}_i = 0$, $T_L$ has no operationally defined meaning. It is the purpose of the following considerations to show which parameters can be determined by which experiments.

## III. DETERMINATION OF THE PARAMETERS

### A. The Michelson Morley Experiment[10)]

The possible anisotropy of the Two-Way-Lightvelocity (TWL)[11)] with respect to the measuring rods of an inertial system can be detected by means of an experiment of the Michelson-Morley type. By means of an interferometer the difference in optical paths for the TWL is measured along two orthogonal rods of lenghts L and l, if the whole apparatus is rotated about 90°. (see Fig. 1)

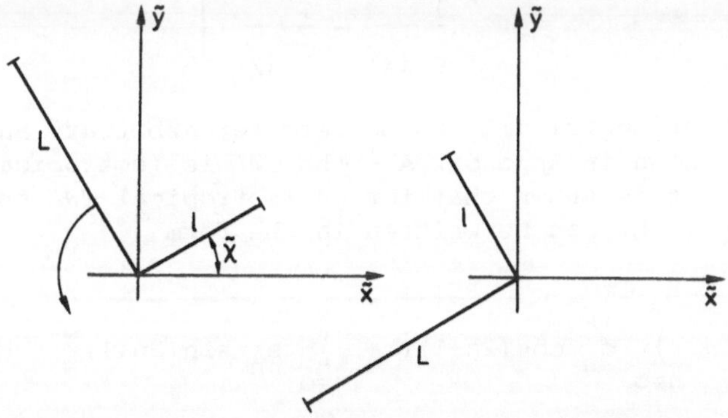


Fig. 1 Arrangement of the measuring rods for the Michelson-Morley experiment.

The most general formula for the velocity vector of the light propagation in an inertial frame is:

$$\tilde{\vec{c}} = \tilde{c}(\vec{r}_o, \tilde{\chi}, \tilde{t})(\cos\tilde{\chi}, \sin\tilde{\chi})$$

where $\vec{r}_o$ denotes the position of the source and $\tilde{\chi}$ the angle between the lightray and the unit vector in x-direction. We will assume that this most general formula for the rectilinear propagation of light has already been reduced by using the following empirical results: The velocity of light neither depends on the position of the source, nor on time.

Thus $\tilde{\vec{c}} = \tilde{c}(\tilde{\chi})(\cos\tilde{\chi}, \sin\tilde{\chi})$ denotes the - possibly anisotropic - velocity of light (with respect to an arbitrary time coordinate $\tilde{t}$) of a source at rest in an inertial frame I. For the times along the measuring rods we get

$$\Delta t_l = l\left(\frac{1}{\tilde{c}^+(\chi)} - \frac{1}{\tilde{c}^-(\chi)}\right), \quad \Delta t_L = L\left(\frac{1}{\tilde{c}^+(\tilde{\chi} + \frac{\pi}{2})} - \frac{1}{\tilde{c}^-(\tilde{\chi} + \frac{\pi}{2})}\right) \qquad \text{(III.1)}$$

and for the difference in optical paths

$$\Delta_o := \Delta t_L - \Delta t_l = L\left(\frac{1}{\tilde{c}^+(\tilde{\chi} + \frac{\pi}{2})} - \frac{1}{\tilde{c}^-(\tilde{\chi} + \frac{\pi}{2})}\right) - l\left(\frac{1}{c^+(\tilde{\chi})} - \frac{1}{c^-(\tilde{\chi})}\right). \qquad \text{(III.2)}$$

After the whole apparatus is rotated about an angle of $90^o$ we get correspondingly

$$\Delta_{q_o} := L\left(\frac{1}{\tilde{c}^+(\tilde{\chi} + \pi)} - \frac{1}{\tilde{c}^-(\tilde{\chi} + \pi)}\right) - l\left(\frac{1}{\tilde{c}^+(\tilde{\chi} + \frac{\pi}{2})} - \frac{1}{\tilde{c}^-(\tilde{\chi} + \frac{\pi}{2})}\right) \qquad \text{(III.3)}$$

[12)]

and for difference of the differences

$$\Delta_{MM} := \Delta_{q_o} - \Delta_o = -(L+l)\left[\left(\frac{1}{\tilde{c}^+(\tilde{\chi} + \frac{\pi}{2})} - \frac{1}{\tilde{c}^-(\tilde{\chi} + \frac{\pi}{2})}\right) - \left(\frac{1}{\tilde{c}^+(\tilde{\chi})} - \frac{1}{\tilde{c}^-(\tilde{\chi})}\right)\right]. \qquad \text{(III.4)}$$

If $\Delta_{MM}$ turns out empirically to be zero for arbitrary angles $\tilde{\chi}$, then - as is shown in Appendix A - the TWL is isotropical in I. In appendix B it is shown that for an isotropical TWL the <u>One-Way-Lightvelocity</u> (OWL) can be written in the form

$$|\tilde{\vec{c}}| = \frac{\tilde{c}}{1 + \sum_{n=o}^{\infty} d_n \cos(2n+1)\tilde{\chi} + \sum_{n=o}^{\infty} e_n \sin(2n+1)\tilde{\chi}} \qquad \text{(III.5)}$$

with arbitrary constants $d_n$ and $e_n$, $\tilde{c}$ being the value of the TWL.

The particular values $|\vec{\tilde{c}}|$ of the OWL in various directions depend only on the way the synchronization procedure for distant clocks has been defined. However, while any given set $(d_n, e_n)$ for $n = 0,1,2,\ldots,$ leads to unique OWLs, no finite set of measurements of the OWL determines all $d_i, e_i$ uniquely. Yet all solutions of (III.5) are equivalent as far as the isotropy of the TWL is concerned. We therefore choose the particular set of parameters $d_i = e_i = 0$ for all $i \in \mathbb{N} \cup \{0\}$. From a physical point of view this choice is equivalent to the choice of one particular clock synchronization: Distant clocks have to be synchronized such that the OWL of a source at rest is isotropic. And this is equivalent to the choice of one particular space-time coordinate system $K(x,y,t)$ with respect to which light propagates isotropically:

$$\vec{c}_L = c_L(\cos\chi,\ \sin\chi). \tag{III.6}$$

Thus the time coordinate $t$ of $K(x^k,t)$ in I has been defined by means of a real physical process, the propagation of light in inertial frames. Conceptually this definition contains two different components, one empirical and one conventional. Empirical is the statement that the TWL of a source at rest in an inertial frame is isotropic, conventional is the agreement to synchronize the clocks such that the OWL is isotropic, as well.[13)]

Repeating the procedure within another inertial frame I' we obtain the equation

$$\vec{\tilde{c}}_{L'} = \tilde{c}'_{L'}(\cos\tilde{\chi}',\sin\tilde{\chi}) \tag{III.7}$$

for the velocity vector of light propagation with respect to the particular coordinate system $\tilde{K}'(\tilde{x}',\tilde{y}',\tilde{t}')$.

In this context it is important to mention that without further assumptions it is not possible to deduce any information about the transformation between different inertial frames from the result of MM experiments performed with sources at rest relative to the measuring apparatus. This is an interesting result since the first MM experiments were performed with resting sources. The isotropy of the TWL is compatible with e.g. the Galileo-transformation. (It is not compatible with the assumption of an ether frame, of course).[14)]

1) The Source is at Rest in $I_o$, the Measurement Takes Place in I'[15)]

Let the propagation of a lightsource at rest in an inertial frame $I_o$ in the coordinates $x,y,t$ be given by the equation

$$\dot{x}^2 + \dot{y}^2 = c^2 \text{ or } \vec{c} = c(\cos\chi,\sin\chi) \tag{III.8}$$

In I' with the coordinate system $\tilde{K}'(\tilde{x}',\tilde{y}',\tilde{t}')$ the propagation of this source reads:

$$\vec{\tilde{c}}'_L(\tilde{\chi}') = \frac{c\, f^1_1 f^2_2}{f^o_o\sqrt{(f^2_2)^2\cos^2\tilde{\chi}' + (f^1_1)^2\sin^2\tilde{\chi}'} + f^o_1 f^2_2\cos\tilde{\chi}' + f^o_2 f^1_1\sin\tilde{\chi}'}(\cos\tilde{\chi}', \sin\tilde{\chi}') \quad \text{(III.9)}$$

And the difference of the optical paths as given by a MM experiment is

$$_L\tilde{\Delta}'_{MM} = -(\tilde{L}'+\tilde{l}')\left[\left(\frac{1}{\tilde{c}'^{+}_L(\tilde{\chi}'+\frac{\pi}{2})} - \frac{1}{\tilde{c}'^{-}_L(\tilde{\chi}'+\frac{\pi}{2})}\right) - \left(\frac{1}{\tilde{c}^{+}_L(\tilde{\chi}')} - \frac{1}{\tilde{c}^{-}_L(\tilde{\chi}')}\right)\right] \quad \text{(III.10)}$$

$$= -(\tilde{L}'+\tilde{l}')\frac{2f^o_o}{f^1_1 f^2_2}\left[\sqrt{(f^2_2)^2\sin^2\tilde{\chi}' + (f^1_1)^2\cos^2\tilde{\chi}'} + \sqrt{(f^2_2)^2\cos^2\tilde{\chi}' + (f^1_1)^2\sin^2\tilde{\chi}'}\right]$$

Due to the linear independence of cos and sin and the initial conditions $f^1{}_1(0) = f^2{}_2(0) = 1$ we get for $_L\tilde{\Delta}'_{MM} = 0$

$$f^1{}_1(v) = f^2{}_2(v) \quad \text{(III.11)}$$

for all v. Inserting III.11 into III.9 we get the simpler expression

$$\vec{\tilde{c}}'_L = \frac{c\, f^1_1}{f^o_o + f^o_1\cos\tilde{\chi}' + f^o_2\sin\tilde{\chi}'}(\cos\tilde{\chi}', \sin\tilde{\chi}'). \quad \text{(III.9')}$$

Geometrically this equation describes rotated ellipses in the x'y'-plane which have one focus in the origin of the coordinate system (see Fig. 2).

The angle of rotation is $\tilde{\chi}'_o = \mathrm{arctg}(f^o_2/f^o_1)$. The source L whose OWL is isotropic in $I_o$ need not have isotropic propagation in I'.

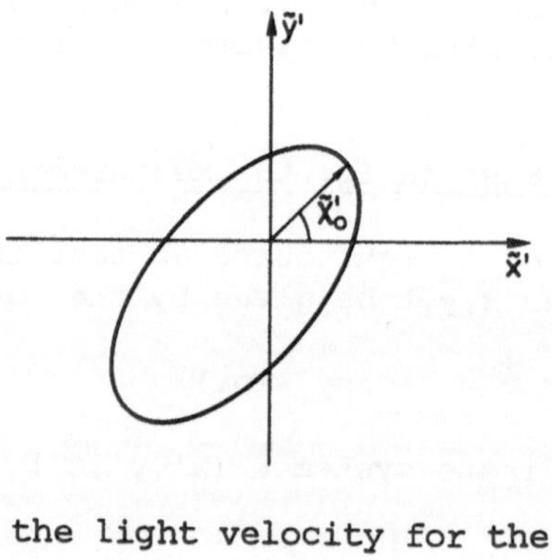


Fig. 2 The vector of the light velocity for the moving source in I'.

We summarize:
In I' the isotropy of the TWL is independent of the motion of the source.

B. The Kennedy-Thorndike Experiment[16)]

As far as the principal procedure of measuring optical paths along two different orthogonal rods is concerned, the KT experiment is comparable to the MM experiment. However, whereas the exact equality of the two rods was of contingent nature for the MM experiment, the difference in lenghts is now important. The apparatus is not rotated, instead the same apparatus is considered in different frames of reference. A systematic analysis, based exclusively on processes in inertial frames, forbids the use of the same measuring device within different inertial frames. However, we can instead assume instead that the KT experiment is performed in one inertial frame with different - resting and moving - sources. In addition we will assume that the empirical results are the same.

1) The Source is at Rest in $I_o$, the Measurement takes Place in I'

Deviating from the procedure of the MM experiment, the difference in optical paths is now measured directly. For a given angle $\tilde{\chi}'$ we get

$$ {}_L\tilde{\Delta}'_{KT} = \frac{\tilde{L}'}{\tilde{c}'_{TW}(\tilde{\chi}' + \frac{\pi}{2})} - \frac{\tilde{l}'}{\tilde{c}'_{TW}(\tilde{\chi}')} $$

$$ = \frac{2f^o_o}{c f^1_1 f^2_2}\left[\tilde{L}'\sqrt{(f^2_2)^2\sin^2\tilde{\chi}' + (f^1_1)^2\cos^2\tilde{\chi}'} - \right. \qquad \text{(III.12)} $$

$$ \left. - \tilde{l}'\sqrt{(f^2_2)^2\cos^2\tilde{\chi}' + (\sin^2\tilde{\chi}'}\right]. $$

Using the results of section A we get

$$ {}_L\tilde{\Delta}'_{KT} = \frac{2(\tilde{L}' - \tilde{l}')}{c}\,\frac{f^o_o}{f^1_1}. \qquad \text{(III.13)} $$

Both $f^o_o$ and $f^1_1$ can depend on v. Without further assumptions about the validity of the RP no statements can be made about the relations between the measuring rods L,l and $\tilde{L}'$, $\tilde{l}'$ respectively. ($\tilde{L}'$ and $\tilde{l}'$ might also depend on v). Therefore unlike Robertson, we cannot directly deduce restrictions for $f^o_o$ and $f^1_1$ from the above results[17)].

2) The Source is at Rest in I', the Measurement takes Place in I'

For a source at rest in I' we get for the difference in optical paths

$$_{L'}\tilde{\Delta}'_{KT} = \frac{2(\tilde{L}'-\tilde{l}')}{\tilde{c}'_{L'}} \tag{III.14}$$

where $\tilde{c}'_{L'}$ is the OWL of the source L' in I'. For $_{L}\tilde{\Delta}'_{KT} = {}_{l'}\tilde{\Delta}'_{KT}$ we have

$$\frac{f^o_o}{f^1_1} = \frac{c}{c'_{L'}} \tag{III.15}$$

From III.9' we already know

$$\tilde{c}'_L = \frac{cf^1_1}{f^o_o} , \tag{III.16}$$

hence we get

$$\tilde{c}'_{L'} = \tilde{c}'_L \quad . \tag{III.16'}$$

We summarize:
The numerical value of the TWL in I' is independent of the motion of the lightsource.

## C. The Ives-Stilwell Experiment[18)]

By means of the IS experiment the time dilatation as predicted by the Special Theory of Relativity can be tested empirically.

A transmitter in motion relative to the receiver emits light signals in forweard and backward direction of the motion. These signals meet the receiver with Doppler-shifts. The measuring device can register these signals simultaneously such that the linear Doppler shift can be eliminated and the quadratic Doppler shift (the time-dilation) can be observed directly.

a) Transmitter rests in $I_o$, Receiver rests in I'

Let the experimental conditions be such that the origin of $\tilde{K}'$ in I' first moves towards the origin of K in $I_o$, coincides with it and then moves away from it. Since no rotation of the apparatus is necessary for the performance of this experiment, we restrict ourselves to the one-dimensional problem and choose the correspon-ding x-axes as directions of the motions. If a light source L in $I_o$, coupled with a light clock (see Fig. 3), emits signals separated by the constant temporal distance $\Delta t_o$, then these signals will be registered in $\tilde{K}'(\tilde{x}',\tilde{t}')$ of I' with the intervals

$$\Delta\tilde{t}^{+'} = \frac{(f^o_o - f^o_1)(1 - \frac{v}{c})\Delta t_o}{\sqrt{1 - \frac{v^2}{c^2}}} \qquad \text{(III.17)}$$

or

$$\Delta\tilde{t}^{-1} = \frac{(f^o_o - f^o_1)(1 + \frac{v}{c})\Delta t_o}{\sqrt{1 - \frac{v^2}{c^2}}} \qquad \text{(III.18)}$$

depending on whether the detector in I' is moving towards the source or away from it.

As a dimensionless parameter we use the linear Doppler shift/ $\Delta\tilde{t}_{o'}$

$$\Delta'_D := \frac{\Delta t'^- - \Delta t'^+}{2\,\Delta t'_o} = \frac{(f^o_o + f^o_1)\frac{v}{c}}{\sqrt{1 - \frac{v^2}{c_2}}} \; \frac{\Delta t_o}{\Delta t'_o} \; . \qquad \text{(III.19)}$$

In linear approximation this represents the velocity v/c.

$$\tilde{\Delta}'_D \simeq \frac{v}{c} \qquad \text{(III.20)}$$

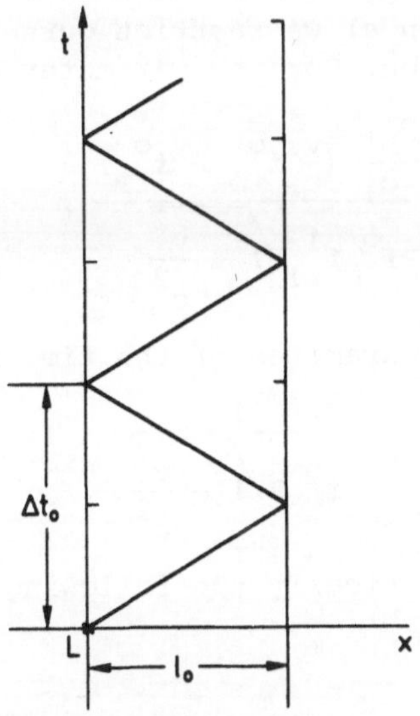


Fig. 3 A light clock in $I_o$.

$\Delta\tilde{t}_o'$ denotes the period of a light clock in I' which is built up in the same way like the emitter clock. This means: If in $I_o$ a light clock consists of a light source L and two opposite mirrors at the distance $l_o$, then in I' a lightclock consists of a source L' and two mirrors at the distance $\tilde{l}_o'$. However, since the connexion between $l_o$ and $\tilde{l}_o'$ is not known, no conclusions can be drawn concerning the connexion between $\Delta t_o$ and $\Delta\tilde{t}_o'$. Thus $F(v) := \Delta t_o/\Delta\tilde{t}_o'$ has to be considered as an additional function of v (with the initial condition F(O) = 1) which has to be determined by the experiments. We will assume that due to the spatial isotropy, the ratio of measuring rods might depend on the magnitude of the velocity but not on its direction, i.e. F is an even function of v.

For the corresponding parameter of time dilation in I' we get

$$\Delta_t' := \frac{\Delta t^{+'} + \Delta t^{-1}}{2\Delta t_o'} - 1 = \frac{(f^o{}_o + f^o{}_1)}{\sqrt{1 - \frac{v^2}{c^2}}} - 1. \qquad \text{(III.21)}$$

Expanding III.21 in terms of $\tilde{\Delta}_{D'}$ to second order gives

$$\tilde{\Delta}_t' = f^{o'}{}_1 \Big|_o \tilde{\Delta}_D' + \frac{1}{2}\left[ (Ff^o{}_o)'' \Big|_o + 1 \right] \Delta_D'^2 \qquad \text{(III.22)}$$

Here the primes at the functions mean derivative with respect to $\tilde{\Delta}_{D'}$.

b) Transmitter rests in I', Receiver rests in $I_o$

Like in subsection a) we restrict ourselves to the analysis of one spatial dimension. For the parameter of the Doppler shift we now get

$$\Delta_D = \frac{\Delta t^+ - \Delta t^-}{2\Delta t_o} = \frac{\frac{c_{L'}'}{c}\left(\frac{v}{c} f^o{}_o - f^o{}_1\right)}{Ff^o{}_o f^1{}_1 \sqrt{1 - \frac{v^2}{c^2}}}, \qquad \text{(III.23)}$$

and the corresponding parameter of the time dilation reads

$$\Delta_t = \frac{\Delta t^+ + \Delta t^-}{2\Delta t_o} - 1 = \frac{1}{f^o{}_o F\sqrt{1 - \frac{v^2}{c^2}}} - 1. \qquad \text{(III.24)}$$

Let us assume that empirically the following functions have been determined

$$\Delta_t = \frac{1}{2} B \, \Delta_D^2 \qquad \text{(III.25a)}$$

$$\tilde{\Delta}'_t = \tilde{A}\,\tilde{\Delta}'_D + \frac{1}{2}\tilde{B}\;\tilde{\Delta}'^2_D \qquad \text{(III.25b)}$$

with particular values for $B,\tilde{A},\tilde{B}$. Since $\Delta_t$ is an even function no linear term can occur in III.25a. By means of an expansion it can easily be shown that the parameters $f^o{}_1$, $f^o{}_o$ and the function F are connected with the values $B,\tilde{A},\tilde{B}$ by the following relations

$$f^o{}_1 = \tilde{A}\,\frac{v}{c} + o\left(\frac{v}{c}\right)^3, \quad Ff^o{}_o = 1 + \frac{1}{2}(B-1)\,\frac{v^2}{c^2} + o\left(\frac{v^4}{c^4}\right). \qquad \text{(III.26)}$$

And since our coordinate systems are chosen such that light emitted from a source at rest in an inertial frame propagates isotropically, an additional relation between $\tilde{A},\tilde{B},B$ is given

$$2 - \tilde{B} = B(1 - \tilde{A})^2. \qquad \text{(III.27)}$$

What do these results tell us about the validity of the PCL or the RP?

In a weak form the PCL says that the equation of propagation for a light source at rest in an inertial system is the same as the equation of propagation for a moving source. ($PCL_W$)

From III.7 we get for the light velocity of L' in two opposite directions in I'

$$\tilde{c}'^{\pm}_{L'} = \pm\,\tilde{c}'_{L'}\,, \qquad \text{(III.28)}$$

and from III.9' we can immediately see what the equivalent values for a source L look like:

$$c^{\pm}_L = \pm\,\frac{cf^1{}_1/f^o{}_o}{1 \pm f^o{}_1/f^o{}_o}$$

The equivalence of the TWLs is already known from III.16'. If, in addition, $f^o{}_1 = 0$ can be confirmed empirically, then $PCL_W$ is satisfied. Indeed no linear effect in III.25b has been observed. This means $\tilde{A} = 0$ and via III.26: $f^o{}_1 = 0$. An analogous analysis of the IS experiment for motions along the y-axes yields the result $f^o{}_2 = 0$.

Thus the results of the IS experiment together with those of the MM and KT experiments allow us to make statements about the validity of the PCL.

Obviously the numerical values of B and $\tilde{B}$ (provided $B + \tilde{B} = 2$ is fulfilled) are arbitrary as far as the PCL alone is concerned. However, the empirically determined values are $B = \tilde{B} = 1$.

In this case the two inertial frames $I_o$ and I' are indistinguishable as far as the propagation of light is concerned: In both frames light propagates isotropically, and the propagation is independent of the motion of the source. Only the numerical values of the lightvelocity can differ from one inertial frame to the next: In $I_o$ this value was fixed by the choice of the coordinate system: $dx/dt_{Light} =: c$. Accordingly we have in I': $d\tilde{x}'/d\tilde{t}' = \tilde{c}_{L'} = f^1_{\ 1}/f^o_{\ o}c$. In order to get a concrete relation between the light velocities in different frames we have to determine the functions $f^i_{\ i}$. It is the indistinguishability of the frames which enables us to determine the remaining parameters.

D. Determination of the Scale Parameters

As far as the propagation of light is concerned, the two inertial frames $I_o$ and I' are empirically equivalent. Thus, instead of starting from the inertial frame $I_o$ with the coordinate system K(x,y,t) and the transformation to I'

$$\begin{aligned} x \rightarrow x' &= \frac{x - vt}{\sqrt{1 - \frac{v^2}{c^2}}} &\longrightarrow\quad & x' = f^1_{\ 1}\tilde{x}' \\ y \rightarrow y' &= y &\longrightarrow\quad & y' = f^1_{\ 1}\tilde{y}' \\ t \rightarrow t' &= \frac{t - \frac{vx}{c^2}}{\sqrt{1 - \frac{v^2}{c^2}}} &\longrightarrow\quad & t' = f^o_{\ o}\tilde{t}' \end{aligned} \tag{III.30}$$

we could also start from the inertial frame I' with the coordinate system $\tilde{K}'(\tilde{x}',\tilde{y}',\tilde{t}')$ and the transformation

$$\begin{aligned} \tilde{x}' \rightarrow \tilde{x} &= \frac{\tilde{x}' - \tilde{v}'\tilde{t}'}{\sqrt{1 - \frac{\tilde{v}'^2}{\tilde{c}'^2}}} &\longrightarrow\quad & x = f^1_{\ 1}(\tilde{v}')\,\tilde{x} \\ \tilde{y}' \rightarrow \tilde{y} &= \tilde{y}' &\longrightarrow\quad & y = f^1_{\ 1}(\tilde{v}')\,\tilde{y} \\ \tilde{t}' \rightarrow \tilde{t} &= \frac{\tilde{t}' - \frac{\tilde{v}'\tilde{x}'}{\tilde{c}'^2}}{\sqrt{1 - \frac{\tilde{v}'^2}{c'^2}}} &\longrightarrow\quad & t = f^o_{\ o}(\tilde{v}')\tilde{t} \end{aligned} \tag{III.31}$$

Since $I_o$ and I' are indistinguishable, the functions $f^i_{\ i}$ in III.31 are the same as in III.30, and $\tilde{v}'$ is the velocity of $I_o$ with respect to I' in the coordinates of I'. For the velocities v and $\tilde{v}'$ the following relations hold

$$\left.\frac{dx}{dt}\right|_{d\tilde{x}'^k=0} =: v = -\frac{f^1{}_1(\tilde{v}')}{f^o{}_o(\tilde{v}')}\,\tilde{v}' =: u(\tilde{v}') \qquad \text{(III.32)}$$

and

$$\left.\frac{d\tilde{x}'}{d\tilde{t}'}\right|_{dx^k=0} =: \tilde{v} = -\frac{f^1{}_1(v)}{f^o{}_o(v)}\,v = u(v). \qquad \text{(III.33)}$$

From the equations (III.32) and (III.33) we get the functional equation

$$u(u(v)) = v, \qquad \text{(III.34)}$$

which due to the antisymmetrie of u has only the two solutions

$$u(v) = +v \quad \text{and} \quad u(v) = -v. \qquad \text{(III.35)}$$

Since $u(v) = -vf^1{}_1(v)/f^o{}_o(v)$ and $f^i{}_i$ are even functions of v, we have in the limit $v \to 0$

$$u(v) \simeq -v$$

The initial condition thus eliminates one of the concurring solutions and we get

$$u(v) = -v = \tilde{v}'. \qquad \text{(III.36)}$$

If I' moves relative to $I_o$ with the velocity v, then $I_o$ moves relative to I' with the velocity -v. This property is usually deduced from group theoretical arguments, here it follows from the indistinguishability of the two inertial frames with respect to the light propagation[19].

From III.33 we now find

$$f^o{}_o(v) = f^1{}_1(v), \qquad \text{(III.37)}$$

and since $-\tilde{v}' = v$ implies $|\tilde{c}'| = c$ we get the PCL in the strong form ($PCL_s$), saying that the <u>numerical value</u> of the light velocity is the same in all inertial frames.

Inserting (III.37) into III.30 and III.31 and combining the two sets of transformations, we find:

$$\tilde{y}' = f^o{}_o(v)y' = f^o{}_o(v)y = f^o{}_o(v)f^o{}_o(-v)\tilde{y} = f^o{}_o{}^2(v)\tilde{y}', \qquad \text{(III.38)}$$

i.e. $f^o{}_o{}^2(v) = 1$, and the initial condition leaves us with one solution

$$f^o{}_o(v) = 1. \qquad \text{(III.39)}$$

This completes our derivation of the Lorentz-transformation.
Once $f^o_o$ is known we can determine F from section C:

$$F(v) = 1.$$

This knowledge enables us to use equivalent measuring rods and clocks in different inertial frames, it is this knowledge which makes a comparison of measuring devices and results in different frames of reference possible.

## IV. SUMMARY AND DISCUSSION

The results of the three experiments MM, KT and IS, performed in different inertial frames, uniquely determine the Lorentz-transformation to second order in v/c. Apart from geometrical suppositions no further assumptions have been made concerning the validity of either the RP or the PCL. About the two principles the three experiments teach us the following:

- the isotropy of the TWL is independent of the motion of the source (MM)
- the numerical value of the TWL is independent of the motion of the source (KT)
- the numerical value of the OWL is independent of the motion of the source.

Thus the conditions of the weak PCL are satisfied, which says that the equation of propagation for a light source at rest is the same as the propagation equation of a moving source. In addition, since inertial frames are indistinguishable with respect to the empirical results of the three experiments, the conditions of the strong PCL are satisfied as well. The strong PCL says that the numerical value of the light t velocity is the same in all inertial frames. Furthermore, this indistinguishability leads to a unique determination of the Lorentz-transformation to second order in v/c. Thus, not only the requirements of the PCL but in addition also those of the RP are satisfied.

The above considerations demonstrate that light sources and light clocks as physical processes satisfy the requirements of both the RP and the PCL. Once this is known measuring rods as defined above and clocks equivalent to light clocks can be used to determine empirically whether other physical processes meet the requirements of the RP or the PCL or both.

## Appendix A

If $\Delta_{MM} = 0$, then the TWL is isotropic.

Proof

Let $\tau$ be an arbitrary angle $\tau \in (0^{o}, 90^{o})$, $\Delta_o$ the difference in optical paths for an angle $\chi$, $\Delta_\tau$ the corresponding difference for the angle $\tau$. (cf. Fig. 1).
Then

$$\Delta^{\tau}_{MM} =: \Delta_\tau - \Delta_o = L\left(\frac{1}{c_{TW}(\chi+\frac{\pi}{2}+\tau)} - \frac{1}{c_{TW}(\chi+\frac{\pi}{2})}\right) - l\left(\frac{1}{c_{TW}(\chi+\tau)} - \frac{1}{c_{TW}(\chi)}\right)$$

$$= \left\{\frac{L}{c_{TW}(\chi+\frac{\pi}{2}+\tau)} - \frac{-l}{c_{TW}(\chi+\tau)}\right\} - \left\{\frac{L}{c_{TW}(\chi+\frac{\pi}{2})} - \frac{l}{c_{TW}(\chi)}\right\}$$

$$=: f(\chi+\tau) - f(\chi) = 0$$

Since $\tau$ is arbitrary, f is constant .
Thus

$$\frac{L}{c_{TW}(\chi+\frac{\pi}{2})} - \frac{l}{c_{TW}(\chi)} = \mathrm{const}_{(\chi)}$$

L and l are assumed to be independent, therefore the TWL does not depend on $\chi$[20].

$$c_{TW}(\chi) = \mathrm{const.}$$

or equivalently

$$\frac{1}{c^{+}(\chi)} - \frac{1}{c^{-}(\chi)} = \mathrm{const.}$$

Appendix B

Obviously for any $\chi$ the relation $c^{-}(\chi) = -c^{+}(\chi + \pi)$ holds. Thus formula A2 can be written as

$$\frac{1}{c^{+}(\chi)} + \frac{1}{c^{+}(\chi + \pi)} = \mathrm{const.} \qquad (B1)$$

or

$$\tilde{f}(\chi) + \tilde{f}(\chi + \pi) = \mathrm{const} =: 2\, k_o \qquad (B2)$$

with $\tilde{f}(\chi) := 1/c^{+}(\chi)$.
Fourier analysis shows that the most general solution of this equation is

$$f(\chi) = k_o + \sum_{n=o}^{\infty} (d_n \cos(2n+1)\chi + e_n \sin(2n+1)\chi). \qquad (B3)$$

NOTES

1 See A. Einstein [1905]. For the derivation of the generalized Lorentz-transformation see W.v. Ignatowski [1910a],[1911a,b]; Ph. Franck and H. Rothe [1911], [1912]; More modern derivations of this transformations can be found in Y.P. Terletzkii [1968], p. 17; G. Süßmann [1969]; P. Mittelstaedt [1980], p. 178 ff; J.-M. Lévy-Leblond [1976].

2 H.P. Robertson [1949], see H.P. Robertson and T.W. Noonan 1968 , p. 69 ff.

3 A frame of reference in which the reference points do not move relative to each other is called a rigid frame of reference.

4 For detailed discussions of these notions see e.g. P. Mittelstaedt [1980] p. 35 ff.

5 This can be done, e.g., by synchronizing all local clocks along the trajectory of a test particle such that the velocity of the uninfluenced test particle, as measured by these clocks, is uniform.
In addition, we shall choose spatial coordinates such that the motion of these test particles are described by linear equations. These are the so called rectilinear coordinates, and since our physical space is assumed to be Euclidean, we adopt the particular choice of Cartesian coordinates.

6 The linear transformations are not the most general transformations which transform straight lines into straight lines. These are the so called collineations or projections. The requirements of homogeneity of space and time reduce these collineations to linear transformations. For details see J. Pfarr [1976] p. 310.

7 The constant c here denotes only a scale factor with the dimension of a velocity. c has no physical meaning in this equation.

8 A detailed discussion of the transformation II.6 can be found in Y.P. Terletzkii [1968] p. 20 ff; cf. P. Mittelstaedt [1980] p. 178 ff. For geometrical reasons Terletzkii dropped the y- and z-components for the time coordinate. Since here the temporal coordinate will be defined by means of a synchronization procedure the two additional components are necessary.

9 c denotes the vacuum velocity of light. Its introduction at this point is of mere technical significance. The operational meaning of c will be defined later.

10 The experiments by Michelson and Morley [1887] have been discussed in the literature in detail. We therefore restrict ourselves to the discussion of basic properties. A review of the experiments of the Michelson-Morley type which have been performed in the past can be found in W.K.H. Panofsky and M. Phillips [1962], p. 277.

11 The TWL is defined by

$$c_{TW} := \frac{2c^+c^-}{c^- - c^+}$$

where $c^+$ and $c^-$ denote the velocities of light in two opposite directions.

12 Obviously the exact equality of the measuring rods L and l is of no significance as far as the result of the experiment is concerned.

13 A detailed analysis of the clock synchronization procedure according to Einstein [1905] and its meaning for the definition of a time coordinate t is given in P. Mittelstaedt [1980] p. 44.

14 For details see J. Pfarr [1981].

15 MM experiments with light emitted by sources in motion with respect to the measuring device have indeed been performed. Tomaschek [1924] used light from the sun, from the moon and from several stars for his experiments.

16 R.J. Kennedy and E.M. Thorndike [1932].

17 From the statement that ${}_L\Delta'_{KT}$ is independent of the velocity of the frames of reference, Robertson [1949] can already conclude that $f^1{}_1/f^o{}_o$ is independent of the velocity.

18 H.E. Ives and G.R. Stilwell [1938], [1941].

19 The reciprocity principle is discussed in detail by V. Berzi and V. Gorini [1969].

20 Note that for L = l only the conclusion can be drawn that the TWL is periodic with period $\pi/2$. In order to confirm the isotropy of the TWL, the MM experiment would have to be performed at least twice with measuring rods of different lengths.

## BIBLIOGRAPHY


Berzi, V. and V. Gorini, J. Math. Phys. 10 (1969) 1518.

Einstein, A. (1905), Ann. d. Phys. 17 (1905) 891.

Franck, Ph. und H. Rothe, Ann. Phys. 34 (1911) 825; Phys. Zeitschr. 13 (1912) 750.

Grünbaum, A., Philosophical Problems of Space and Time, 2nd ed., D. Reidel Publ. Co., Dordrecht-Holland 1973.

Ignatovski, W. v., Arch. Math. Phys. 17 (1910) 1(a); 18 (1911) 17(a); Phys. Zeitschr. 11 (1910) 972(b); 12 (1911) 779(b).

Ives, H.E. and G.R. Stilwell, J. Opt. Soc. Am. 28 (1938) 215, 31 (1941) 369.

Kennedy, R.J. and E.M. Thorndike, Phys. Rev. 42 (1932) 400.

Mansouri, R. und R.U. Sexl, GRG 8 (1977) 497, 515, 809.

Lévy-Leblond, J.-M., Am. J. Phys. 44 (1976) 271.

Michelson, A.A. and E.W. Morley, Am. J. Sci. 34 (1887) 333; Phil. Mag. 24 (1887) 449.

Mittelstaedt, P., Der Zeitbegriff in der Physik, Bibliographisches Institut, Mannheim, 2nd ed. 1980.

Panofsky, W.K.H. and M. Phillips, Classical Electricity and Magnetism, Addison-Wesley, Reading Mass. 1962.

Pfarr, J., Zeitschr. f. all. Wissenschaftstheorie 7 (1976) 298.

---. Zur Interpretation des Michelson-Versuchs, in J. Nitsch,

J. Pfarr und E.-W. Stachow (eds.), Grundlagenprobleme der modernen Physik, Bibliographisches Institut, Mannheim, (1981).

Reichenbach, H., The Philosophy of Space and Time, Dover Publications, New York 1957.

Robertson, H.P., Rev. Mod. Phys. 21 (1949) 378.

Robertson, H.P. and T.W. Noonan, Relativity and Cosmology, Saunders, Philadelphia 1968.

Süßmann, G., Z. Naturforsch. 24 a (1969), 495.

Terletzkii, Y.P., Paradoxes in the Theory of Relativity, Plenum Press, New York 1968.

Tomaschek, R., Ann. Physik 73 (1924) 105.

# GEOMETRICAL STRUCTURES AND THE GRINDING PROCESS FOR THREE PLATES

D.P.L. Castrigiano

Institut für Mathematik der Technischen Universität
München
D-8000 München 2, Arcisstraße 21
Federal Republic of Germany

## ABSTRACT

The *Grinding Process of Three Plates* (GTP), which is discussed in the philosophy of science chiefly by Dingler, Lorenzen and the protophysicists, is defined mathematically within the frame of Riemannian geometry. For this the notions of *geodesic rigidity at a point* which generalizes that of isometric rigidity, and of a *Dingler surface* which describes the surfaces generated by GTP, are introduced. It is proved that GTP can be carried out *if and only if the space has constant curvature*. Furthermore, in this case the generated surfaces are *just the totally geodesic hypersurfaces*.

## I. INTRODUCTION

In practice the *Grinding Process for Three Plates* (GTP) serves, for example, to produce flat optical surfaces. Three plates are grinded in turn against each other until they *fit each other perfectly*. The use of three instead of two plates excludes that the generated surface is spherical.

Because of its theoretical interest, GTP is discussed by Mach, Poincaré and Dingler and more recently by Lorenzen and the advocates of *protophysics* [1]. The well-known problems in the philosophy of science treated in this connexion will not be considered in this paper. This will be done from a physical point of view in a separate paper.

Here we are concerned with the following two questions:
(a) In which spaces is it possible to carry out GTP? (In an arbitrary space the curvature could prevent the grinding process from

coming to an end, or, for lack of movability of the bodies, even from starting.) (b) Which surfaces are generated in case GTP works?

According to a common opinion, which is partly induced by protophysical considerations [2], GTP should work only in Euclidean spaces. *This* would distinguish the Euclidean geometry from others. However, Süßmann [3] points out that GTP works in any Riemannian space of constant curvature. To illustrate this let us follow [3] and consider the two-dimensional sphere. In this case 'bodies' are surfaces on the sphere (evidently these can be moved freely on the sphere) and 'surfaces' are curves in the sphere. The grinding of two bodies against each other yields a 'surface' which is a circular line whose radius is not greater than that of the sphere. Evidently all these 'surfaces' fit each other (and the bodies are freely movable along their common surfaces) if and only if they are pieces of great circles.

Let us show that GTP works in the general case of a space of constant curvature. Since Helmholtz [4] one knows that the spaces of constant curvature are characterized by the free (unrestricted) mobility of rigid bodies. In such a space a body can be moved to every place and put in every position. Thereby it does not undergo any deformation, i.e. all inner distances and angles are preserved. Now let us consider two arbitrary convex bodies part of whose surfaces are totally geodesic hypersurfaces. In spaces of constant curvature totally geodesic hypersurfaces are just those submanifolds which are generated by all geodesics that start from one point and are perpendicular to a fixed direction at that point. They are invariant under parallel transport along curves lying in them. Now if the two bodies come in contact at one point of the totally geodesic parts of their surfaces then the surfaces are identical in some neighbourhood of this point and, furthermore, the bodies are freely movable along their common surface.

Here the following general remark is in place. Our approach to things is such that *we always look at the result of* GTP *and never describe the grinding process itself*. Otherwise we would have to speak about material properties etc.

In the foregoing considerations we started from a totally geodesic piece of the surface assuming that it was generated by GTP. It was proved that it satisfies the definition of GTP, i.e. that all bodies with such surfaces *fit each other perfectly*. However, the following considerations (see (5)) show that the surfaces generated by GTP are indeed totally geodesic.

The mathematical framework we work in is that of Riemannian spaces of class $C^{\infty}$ (the assumption about differentiability can be weakened easily if needed). All definitions and statements are of local nature and do not demand any global properties. We introduce the notion of *geodesic rigidity at a point* which generalizes essentially that of *isometric rigidity* (cf. Helmholtz [4]) and which allows us to consider *deformable* bodies. With this notion a *Dingler surface* will be defined. It is generated just by applying GTP. Then, our main result states: *If GTP can be carried out at each point and in each position then the space has constant curvature and the Dingler surfaces are totally geodesic hypersurfaces.* Taking into account what we already know about GTP in spaces of constant curvature this gives a complete answer to the questions (a) and (b) posed at the beginning.

Let us recall that our considerations are moving within the frame of Riemannian geometry. It is worthwhile asking if this rich structure is actually needed. Perhaps it could be deduced from a poorer one. We think that Freudenthal [5] could serve as a model. There, the starting point is a locally compact topological space. From an epistemological point of view it would be most satisfactory if one could proceed operationally and, as done in Schmidt [6] (see also Mayr [7]),'construct' the space itself.

## II. GEODESIC RIGIDITY AT A POINT AND DINGLER SURFACES

A grinding process presupposes certain properties of bodies. In this context the properties already mentioned are essential:
(a) *Movability*,
(b) *Rigidity*.
In order not to drift into the channels of Helmholtz's argumentation one has to give more general definitions of (a) and (b). Furthermore, the general situation no longer allows us to consider them as *independent* from each other. Indeed, generally the rigidity of a body influences its movability. For instance, a rigid body may be wedged in between the geodesic field lines (this happens if there are not enough local isometries). In other words this means that, in general, movability is made possible just by deformability. Let us illustrate this by explaining the notion of geodesic rigidity at a point.

Imagine a body which is fixed at one point. Then, in Euclidean space, the movability of a rigid body is limited to rotations around varying axes through the fixed point. With respect to an observer who is carried by the body nothing changes. Movability and rigidity do not allow other motions. However, if the body is deformable then more motions may be possible (still using the Lagrangian description): The points of the body move in an arbitrarily continuous manner along the geodesics which pass through the fixed point. Now (in a general Riemannian space) *geodesic rigidity at the fixed point* is present if the body moves infinitesimally at the fixed point as just described. Such motions are called *geodesic at the fixed point*. Formalizing this we get the following

1. DEFINITION

Let M be a Riemannian space and $z \in M$. A $C^{\infty}$-mapping $f : U \to V$, U and V open neighbourhoods of z, is *geodesic in* z, if
(i) $f(z) = z$,
(ii) $F := T_z(f)$, the differential of f at z, is bijective,
(iii) $(\nabla_{X_z}X - F^{-1}\nabla_{FX_z}T(f)\cdot X)$ is parallel to $X_z$ for any vector field X through z.*)

Two more notions are needed for our definition of a surface generated by GTP. From now on M denotes a Riemannian space of dimension $n \geq 3$. Let us consider $z \in M$, H a hypersurface in M with $z \in H$, U and V open neighbourhoods of z in M, $f : U \to V$ a mapping of class $C^{\infty}$ and $v_1,\ldots,v_n$ a basis of $T_z(M)$ such that $v_1,\ldots,v_{n-1}$ is a basis of $T_z(H)$. By definition f is said to *keep the hyperplane* H *invariant* if $f(U \cap H) \subset V \cap H$; if z is a fixed point of f, i.e. $f(z) = z$, f is said to *preserve* respectively to *reverse the orientation of* H *at* z if $\alpha_n > 0$ respectively $\alpha_n < 0$ where

$$T_z(f)v_n = \sum_{i=1}^{n} \alpha_i v_i$$

with $\alpha_i \in \mathbb{R}$. One verifies by computation that $\alpha_n$, and hence sgn $\alpha_n$, does not depend on the choice of basis. G(z,H) denotes the set of all mappings that are geodesic in z and leave H invariant, and $G_o(z,H)$ denotes the subset of G(z,H) consisting of all mappings that, in addition, preserve the orientation of H in z.**)

2. DEFINITION

Let H be a hypersurface in M and $z \in H$. H is said to be generated by GTP or equivalently to be a *Dingler surface* if the following three conditions are satisfied:
(i) H has constant curvature in a neighbourhood of z,
(ii) for every two vectors $v_1, v_2 \in T_z(H)$ there is $f \in G_o(z,H)$ such that $T_z(f)v_1$ is parallel with $v_2$,
(iii) $G(z,H) \setminus G_o(z,H)$ is non-void.

In order to interpret (2) one notes first that the surfaces of bodies are described piecewise by hypersurfaces. (i) takes into account the fact that grinded plates are freely movable along

---

*) $T(f)\cdot X$ is well-defined as a local vector field in a neighbourhood of $f(z)=z$ since, by (ii), f is a local diffeomorphism.

**) G(z,H) forms a group it it is provided with an adequate equivalence relation. $G_o(z,H)$ is then a subgroup of G(z,H). However, we do not need this construction.

their common surface H. With respect to the rigidity of bodies (i) is not very restrictive. For example, in three dimensional flat space the surface still can be curved: The plates are freely movable in all directions like a label on a bottle as long as the Gaussian curvature vanishes. (ii) expresses a weak isotropy of the grinded surface H at the point z: Any two directions at z on the surface of the plate can be brought to coincidence by motions that are compatible with the geodesic rigidity at z of the plate, i.e. by motions of the plate that we called geodesic at z. Finally, (iii) describes the *act of comparison* which is part of GTP. Grinding two plates against a third one, two *equal* surfaces are generated. Subsequently they are brought in contact and compared with each other. This process *realizes the comparison of the grinded surface with itself*. In this sense a Dingler surface coincides with itself. This is expressed by (iii): There is a geodesic motion at z that keeps H invariant and reverses the orientation of H at z.

Now we require that

(D) for every point z and every hyperplane $P \subset T_z(M)$ there is a Dingler surface H such that $z \in H$ and $P = T_z(H)$.

(D) formalizes the idea that the GTP can be carried out at every place and in every position.

Then the following statement holds.

3. THEOREM

A connected Riemannian space M of dimension $\geq 3$ has constant curvature if and only if it has the property (D).

The direction 'M has constant curvature ⇒ (D) holds' is already sketched in the introduction. It suffices to note that in the case of constant curvature the totally geodesic hypersurfaces are Dingler surfaces. Indeed, the former satisfy even much more restrictive requirements than those imposed by (2). For instance, one could substitute the mappings occurring in (2ii) and (2iii) for local isometries. Thus one can say that the GTP is characteristic of spaces of constant curvature *since there is plenty of scope concerning its mathematical description*.

The following chapter serves to prove the rest of the assertion of (3).

## III. PROOFS

M denotes a connected Riemannian space of dimension $\geq 3$.

4. THEOREM

Let (2ii) and (2iii) be valid. Then H is geodesic at z.

*Proof*. We use the notations of [8]. One has to show that the second fundamental form s of H with respect to M vanishes at z. For any vector field X on H through z and any $f \in G(z,H)$ we have

$$s(X_z, X_z) = \perp(\nabla_{X_z} X) = \perp(F^{-1}\nabla_{FX_z} T(f)\cdot X + v_f) =$$

$$\alpha_f(\perp\nabla_{FX_z} T(f)\cdot X) + 0 = \alpha_f s(FX_z, FX_z)$$

where $v_f \in T_z(H)$ parallel with $X_z$, and $\alpha_f > 0$ respectively $\alpha_f < 0$ if $f \in G_0(z,H)$ respectively $f \notin G_0(z,H)$. Because of (2ii) it follows that s at z is either strictly positive or is strictly negative or is zero. Because of (2iii) only the last case occurs. □

Thus Dingler surfaces are geodesic at least at one point. From the theorem [9;20.23.6] due to E. Cartan it follows

5. COROLLARY

If M has constant curvature then the Dingler surfaces are just the totally geodesic hypersurfaces of M.

Taking (2i) into account we have the situation that any hypersurface which is geodesic at one point has constant curvature in a neighbourhood of that point. In particular, the following lemma applies.

6. LEMMA

Let dim M be $\geq 4$ and $z \in M$. If the sectional curvature does not depend on the sections tangent to any hypersurface which is geodesic at z then it is sectionally independent at z.

*Proof*. Let $K_H$ denote the sectional curvature at z with respect to the hypersurface H geodesic at z. Evidently every surface (= two dimensional submanifold) geodesic at z lies in at least one hypersurface H geodesic at z, ant its curvature at z is equal to $K_H$. Thus it remains to show that $K_{H_1} = K_{H_2}$ for any two hypersurfaces $H_1$ and $H_2$ geodesic at z. Let us consider the nontrivial case, i.e. $T_z(H_1) \neq T_z(H_2)$. Then $\dim(T_z(H_1) \cap T_z(H_2)) = n-2 \geq 2$. Therefore, a surface F exists which is geodesic at z and which lies in $H_1 \cap H_2$. Its curvature is equal to $K_{H_1}$ as well as to $K_{H_2}$. □

From the theorem [9;20.23.2] due to Schur it follows immediately.

7. COROLLARY

If dim $M \geq 4$ and (D) holds then M has constant curvature.

In order to complete the proof of (3) the statement of (7) has to be proved for dim M = 3. Because of (5)f. this follows from

8. THEOREM

Let M be three dimensional. If any surface geodesic at one point has constant curvature in a neighbourhood of this point then M has constant curvature.

*Proof*. Let w be a geodesic where $z := w(0)$ and $v := \dot{w}(0)$. Among all planes in $T_{w(t)}(M)$ containing $\dot{w}(t)$ let $P_1(t)$ respectively $P_2(t)$ denote a plane for which the sectional curvature is minimal respectively maximal sectional curvature. Now let us consider a surface F geodesic at z such that $v \in T_z(F)$. Let $\varphi(t)$ be the angle formed by $T_{w(t)}(F)$ and $P_1(t)$ and let $K(t)$ be the sectional curvature at $w(t)$ for the section $T_{w(t)}(F)$. Since without loss of generality $P_1(t) \perp P_2(t)$ it is

$$K(t) = \cos^2\varphi(t)\ K_1(t) + \sin^2\varphi(t)\ K_2(t).$$

Since the curvature of F is constant in a neighbourhood of z and is equal to the corresponding sectional curvature at z it follows from the Lemma a Synge [8] that $t \rightarrow K(t)$ has a local minimum at $t = 0$. Therefore

$$0 = 2\sin\alpha\cos\alpha\ \dot{\varphi}(0)\ (K_2(0)-K_1(0)) + \sin^2\alpha\ \dot{K}_2(0) + \cos^2\alpha\ \dot{K}_1(0)$$

where $\alpha := \varphi(0)$. This holds for all values of the initial angle $\alpha$. Hence $\dot{K}_1(0) = \dot{K}_2(0) = 0$. Since the foregoing considerations are not limited to $z = w(0)$ one concludes that $\dot{K}_1(t) = \dot{K}_2(t) = 0$ for all t, i.e. $K_1$ and $K_2$ are constant along the geodesic w. Therefore the curvature of F, for which $T_z(F) = E_2(0)$, is equal to the sectional curvature ($= K_2$) along w. It follows from the Lemma of Synge [8] that $E_2(0)$ is mapped on $E_2(t)$ by parallel transport. If $K_1 < K_2$ then necessarily $E_1(t) \perp E_2(t)$. This implies that parallel transport also maps $E_1(0)$ on $E_1(t)$ and, therefore, that the sectional curvature does not change. This is trivial in case of $K_1 = K_2$. Thus, summarizing, we see that the sectional curvature does not change under parallel transport along any direction of the section. Choosing coordinates this means: $R_{ijkl;m}\, v^m u^i u^k v^j v^l = 0$ for all $u,v \in \mathbb{R}^3$, where R denotes the Riemann tensor and ';' covariant differentiation. Because of the symmetry properties of R and the Bianchi identity this is equivalent to $R_{ijkl;m}+R_{ilkm;j}+R_{imkj;l}=0$. By this additional symmetry and the dimension 3 it follows that $R_{ijkl;m} = 0$, i.e. M is locally symmetric.

Since our problem is a local one we may assume by [10;Ch.IV, Theorem 5.1] that M is globally symmetric. According to the classification due to Cartan (see [10;Ch.X,§6 Tables IV,V) there are just two types of *irreducible* symmetric spaces of dimension 3

$$SU(2) \text{ and } SL(2,\mathbb{C})/SU(2).$$

The first contains all spaces of positive constant curvature, the second all those of negative constant curvature.

It remains to examine the reducible spaces. Because of the local nature of our problem we may assume, by [10;Ch.IV, Proposition 3.6], that M is simply connected. Then, by [10;Ch.V, Proposition 4.2], M is a product of irreducible factors. In the case of three (one-dimensional) factors M is Euclidean. Thus the case $M = M_1 \times M_2$, where $\dim M_1 = 1$ and $\dim M_2 = 2$, remains. Since a two-dimensional symmetric space has constant curvature, we may choose the metric tensor to be

$$(g_{ij}) = \begin{pmatrix} 1 & 0 & 0 \\ 0 & h^2 & 0 \\ 0 & 0 & h^2 \end{pmatrix}, \text{ where } h^{-1} = 1 + \frac{K}{4}(y^2+z^2), K \neq 0,$$

in some neighbourhood of the origin. This is the Riemannian form of the metric for $M_2$ where K is the curvature of $M_2$. Now, in the following the case $M = M_1 \times M_2$ is ruled out by explicit computations which show that M does not satisfy the premises of (8).

Using the Gaussian equation we compute the curvature of the surface $F \subset M_1 \times M_2$ which is geodesic at the origin and satisfies $T_o(F) = \mathbb{R}e_2 + \mathbb{R}(\alpha e_1 + \beta e_3)$, where $\alpha^2+\beta^2 = 1$, $\beta \neq 0$ and $(e_j)$ is the natural basis of $\mathbb{R}^3$. The notations are essentially the same as in [8]. Primes refer to F.

$$\Gamma^2_{22} = \Gamma^3_{23} = \Gamma^3_{32} = -\Gamma^2_{33} = -\frac{K}{2}hy,$$
$$\Gamma^3_{33} = \Gamma^2_{23} = \Gamma^2_{32} = -\Gamma^3_{22} = -\frac{K}{2}hz,$$

all other Christoffel symbols are zero,

$$R_{2332} = -R_{3232} = R_{3223} = -R_{2323} = K,$$

all other components of the curvature tensor are zero.

*The case* $K > 0$.

The geodesic through O and with tangent vector $u\alpha e_1 + ve_2 + u\beta e_3$, $u^2 + v^2 = 1$, at O, parametrized by arc length, is given by

$$x = u\alpha t,\ y = \frac{v}{c}\sqrt{\frac{4}{K}}\tan\sqrt{\frac{K}{4}}\ ct,$$

$$z = \frac{u\beta}{c}\sqrt{\frac{4}{K}}\tan\sqrt{\frac{K}{4}}\ ct,\ c = \sqrt{1-u^2\alpha^2}.$$

Hence we have the following parametrization of F by $\tau := \sqrt{\frac{K}{4}}\, ct$ and $\theta$, where $\cos\theta = u\beta/c$, $\sin\theta = v/c$:

$$x = \frac{\alpha}{\beta}\sqrt{\frac{4}{K}}\ \tau\cos\theta,\ y = \sqrt{\frac{4}{K}}\tan\tau\sin\theta,$$

$$z = \sqrt{\frac{4}{K}}\tan\tau\cos\theta.$$

With respect to this parametrization we have

$$g'_{11} = \frac{4}{K}\left(1 + \frac{\alpha^2}{\beta^2}\cos^2\theta\right),\ g'_{12} = g'_{21} = -\frac{4}{K}\frac{\alpha^2}{\beta^2}\tau\sin\theta\cos\theta,$$

$$g'_{22} = \frac{4}{K}\left(\frac{\alpha^2}{\beta^2}\tau^2\sin^2\theta + \sin^2\tau\cos^2\tau\right),$$

$$g' = \frac{16}{K^2}\left(\frac{\alpha^2}{\beta^2}\tau^2\sin^2\theta + \frac{\alpha^2}{\beta^2}\sin^2\tau\cos^2\tau\cos^2\theta + \sin^2\tau\cos^2\tau\right),$$

$$\Gamma'^{1}_{11} = \Gamma'^{2}_{11} = 0,$$

$$\Gamma'^{1}_{12} = \Gamma'^{1}_{21} = \frac{16}{K^2}\frac{\alpha^2}{\beta^2}\frac{1}{g'}\left(\frac{1}{4}\tau\sin 4\tau - \sin^2\tau\cos^2\tau\right)\sin\theta\cos\theta,$$

$$\Gamma'^{2}_{12} = \Gamma'^{2}_{21} = \frac{16}{K^2}\frac{1}{g'}\left(\frac{1}{4}\frac{\alpha^2}{\beta^2}\sin 4\tau\cos^2\theta + \frac{\alpha^2}{\beta^2}\tau\sin^2\theta + \frac{1}{4}\sin 4\tau\right),$$

$$\Gamma'^{1}_{22} = -\frac{16}{K^2}\frac{1}{g'}\left(\frac{1}{4}\frac{\alpha^2}{\beta^2}\tau^2\sin 4\tau\sin^2\theta + \frac{\alpha^2}{\beta^2}\tau\sin^2\tau\cos^2\tau + \frac{1}{4}\sin^2\tau\cos^2\tau\sin 4\tau\right),$$

$$\Gamma'^{2}_{22} = \frac{16}{K^2}\frac{\alpha^2}{\beta^2}\frac{1}{g'}\tau\left(\tau - \frac{1}{4}\sin 4\tau\right)\sin\theta\cos\theta$$

(the last two Christoffel symbols are not needed in the sequel). Then the second fundamental form satisfies

$$\Pi_{11} = 0,$$

$$(\Pi_{12})^2 = \frac{4}{K}\frac{\alpha^2}{\beta^2}\left(\frac{16}{K^2}\frac{1}{g'}\right)^2\left\{\frac{1}{16}[\tau\sin 4\tau - \sin^2 2\tau]^2 + \right.$$

$$\frac{\alpha^2}{\beta^2}\left[\tau^4\cos^2 2\tau\sin^2\theta - \tau^3\sin 2\tau\cos 2\tau\sin^2\theta + \right.$$

$$\tau^2 \frac{1}{4} \sin^2 2\tau (1-\sin^2 2\tau \cos^2\theta) -$$

$$\tau \frac{1}{4} \sin^3 2\tau \cos 2\tau \cos^2\theta + \frac{1}{16} \sin^4 2\tau \cos^2\theta \Big]\Big\} \sin^2\theta.$$

Using the Gaussian equation the curvature k of F follows:

$$k = \frac{1}{g'} \left[\frac{4}{K} \sin^2 2\tau - (\Pi_{12})^2\right],$$

$$k(0) = \beta^2 K, \quad \dot{k}(0) = 0, \quad \ddot{k}(0) = -\frac{16}{9} \beta^2 (1-\beta^2) K \sin^2\theta,$$

where the dot indicates differentiation with respect to $\tau$. This means that for $0 < |\beta| < 1$ the curvature of F reaches a strict relative maximum at 0, in particular it is not constant.

*The case* $K < 0$
follows from the case $K > 0$ by analytic continuation to imaginary values of $\tau$. In particular $k(\tau,K) = -k(i\tau,-K), K < 0$, $\tau$ real. Thus, for $0 < |\beta| < 1$, again there is a strict relative maximum of k at 0. □

ACKNOWLEDGEMENT

The author is indebted to Prof. G. Süßmann suggesting the subject of this work. He is also grateful to D. Mayr and W. Satzger for many valuable discussions.

REFERENCES

1. (a) Mach, E., Erkenntnis und Irrtum, 1905, fifth edition,p.269
   (b) Poincaré, H., Science et Méthode, Paris 1908, p.415
   (c) Dingler, H., Die Grundlagen der angewandten Geometrie, Leipzig 1911, p.21f.
   Dingler, H., Das Experiment. Sein Wesen und seine Geschichte; Reinhardt München, 1928, p.59
   (d) Lorenzen, P., Schwemmer, O., Konstruktive Logik, Ethik und Wissenschaftstheorie; B.I.-Hochschultaschenbücher Band 700 Mannheim 1975, p.228
   (e) Böhme, G. (editor), Protophysik; Suhrkamp Frankfurt 1976, in particular P. Janich: Zur Protophysik des Raumes, p.106f.
2. see e.g. [1] (e)
3. Süssmann, G., Physikalische Kennzeichnung der Clifford-Kleinschen Räume; talk at the conference on 'Grundstrukturen einer physikalischen Theorie: Raum, Zeit und Mechanik', Munich May 1979, to be published

4. v. Helmholtz, H., Über die Tatsachen, welche der Geometrie zugrunde liegen; Nachr. d. Ges. d. Wiss. Göttingen 1868
5. Freudenthal, H., Neuere Fassung des Riemann-Helmholtz-Lieschen Raumproblems; Math. Zeitschr. 63, 374-405 (1956)
Freudenthal, H., Das Helmholtz-Liesche Raumproblem bei indefiniter Metrik; Math. Annalen 156, 263-312 (1964)
6. Schmidt, H.-J., Axiomatic Characterization of Physical Geometry; Lecture Notes in Physics 111, Springer-Verlag Berlin 1979
7. Mayr, D., Zur konstruktiv-axiomatischen Charakterisierung der Riemann-Helmholtz-Lieschen Raumgeometrien und der Poincaré-Einstein-Minkowskischen Raumzeitgeometrien durch das Prinzip der Reproduzierbarkeit; Doctoral thesis, Universität München 1979
8. Spivak, M., A Comprehensive Introduction to Differential Geometry; Publish or Perish, Inc. Berkeley 1979, Vol.III, Ch.1
9. Dieudonné, J., Treatise on Analysis; Academic Press New York 1976, Vol.IV
10. Helgason, S., Differential Geometry, Lie Groups, and Symmetric Spaces; Academic Press New York 1978

PARTICIPANTS

W. BALZER
Seminar für Philosophie
Universität München
Ludwigstraße 31/I
8000 MÜNCHEN 2, West Germany

W. BAYER
Fachbereich Physik
Philipps Universität Marburg
Renthof 7
3550 MARBURG, West Germany

D. P. L. CASTRIGIANO
Institut für Mathematik
Technische Universität München
Arcisstraße 21
8000 MÜNCHEN 2, West Germany

R. M. COOKE
Technische Hogeschool Delft
Kanaalweg 2b
2600 GB DELFT, Netherlands

W. DIEDERICH
Fakultät für Philosophie
Universität Bielefeld
Postfach 86 40
4800 BIELEFELD, West Germany

W. GRAFE
Philosophisches Seminar
Universität Göttingen
Nikolausberger Weg 9 C
3400 GÖTTINGEN, West Germany

A. HARTKÄMPER
Fachbereich Physik
Universität Osnabrück
Postfach 44 69
4500 OSNABRÜCK, West Germany

J. HILGEVOORD
Institut voor theoretische Fysica
Valckemierstraat 65
1018 AMSTERDAM, Netherlands

W. HÖRING
Philosophisches Seminar
Universität Tübingen
Alte Burse, Bursagasse 1
7400 TÜBINGEN, West Germany

P. JANICH
Fachb.Gesellschaftwissenschaften
Philipps Universität Marburg
Krummbogen 28
3550 MARBURG, West Germany

A. KAMLAH
Fachbereich 2
Universität Osnabrück
Postfach 44 69
4500 OSNABRÜCK, West Germany

B. KANITSCHNEIDER
Zentrum für Philosophie
Universität Gießen
Otto-Behagel-Str. 10
6300 GIESSEN, West Germany

L. KRÜGER
Fakultät für Philosophie
Universität Bielefeld
Postfach 86 40
4800 BIELEFELD, West Germany

G. LUDWIG
Fachbereich Physik
Philipps Universität Marburg
Renthof 7
3550 MARBURG, West Germany

U. MAJER
Philosophisches Seminar
Universität Göttingen
Nikolausberger Weg 9 C
3400 GÖTTINGEN, West Germany

D. MAYR
Fachbereich Physik
Philipps Universität Marburg
Renthof 7
3550 MARBURG, West Germany

O. MELSHEIMER
Fachbereich Physik
Philipps Universität Marburg
Renthof 7
3550 MARBURG, West Germany

C. U. MOULINES
Instituto de Investigaciones Filosóficas
Torre 1 de Humanidades, 4. Piso
Ciudad Universitaria
MEXICO 20 D.F.

H. NEUMANN
Fachbereich Physik
Philipps Universität Marburg
Renthof 7
3550 MARBURG, West Germany

J. PFARR
Institut für Theoretische Physik
Universität Köln
Zülpicher Straße 77
5000 KÖLN 41, West Germany

E. SCHEIBE
Philosophisches Seminar
Universität Göttingen
Nikolausberger Weg 9 C
3400 GÖTTINGEN, West Germany

H.-J. SCHMIDT
Fachbereich Physik
Universität Osnabrück
Postfach 44 69
4500 OSNABRÜCK, West Germany

E. W. STACHOW
Institut für Theoretische Physik
Universität Köln
Zülpicher Straße 77
5000 KÖLN 41, West Germany

G. SÜSSMANN
Sektion Physik
Universität München
Theresienstraße 37
8000 MÜNCHEN 2, West Germany

R. WERNER
Fachbereich Physik
Universität Osnabrück
Postfach 44 69
4500 OSNABRÜCK, West Germany

INDEX